Agri-Food Management

Agri-Food Management

Edited by **Jamie Hanks**

SYRAWOOD
PUBLISHING HOUSE

New York

Published by Syrawood Publishing House,
750 Third Avenue, 9th Floor,
New York, NY 10017, USA
www.syrawoodpublishinghouse.com

Agri-Food Management
Edited by Jamie Hanks

International Standard Book Number: 978-1-68286-047-2 (Hardback)

Printed in the United States of America.

Contents

Preface

Effective management strategies are highly important for agricultural produce to reach consumers via proper channels. This book studies, analyses and upholds the pillars of agri-food management and its utmost significance in modern times. It evaluates the current practices of agri-food firm management and marketing, organization of agri-food chains, agri-food policy, etc. It is an excellent resource guide for experts and students who are actively engaged in food technology, agri-business management and allied fields. As this field is emerging at a rapid pace, the contents of this book will help the readers understand the modern concepts of the subject.

The information contained in this book is the result of intensive hard work done by researchers in this field. All due efforts have been made to make this book serve as a complete guiding source for students and researchers. The topics in this book have been comprehensively explained to help readers understand the growing trends in the field.

I would like to thank the entire group of writers who made sincere efforts in this book and my family who supported me in my efforts of working on this book. I take this opportunity to thank all those who have been a guiding force throughout my life.

Editor

On the relationship between regional trade agreements and agricultural technology and productivity

Pascal L Ghazalian

Correspondence: pascal.ghazalian@uleth.ca
Department of Economics, University of Lethbridge, 4401 University Drive, Lethbridge, Alberta T1K 3M4, Canada

Abstract

The implementation of a Regional Trade Agreement (RTA) is normally accompanied with a rise in market competition levels, in domestic agricultural markets through increases in imports and in foreign agricultural markets through increases in exports. These effects are expected to induce adjustments in agricultural technology and productivity in the importing and exporting countries. This paper analyzes the implications of these adjustments in the context of Viner's (The Customs Union Issue. Carnegie Endowment for International Peace: New York, NY, 1950) conventional partial equilibrium framework with perfectly elastic foreign supply schedules faced by the importing member country. It also examines these implications in the context of Pomfret's (Review of World Economics, 122(3): 439-465, 1986) extended partial equilibrium framework depicting upward-sloping foreign supply schedules for the importing member country. The analysis underscores important changes and redistributions through the RTA's initial benefits and losses, following the RTA-induced adjustments in agricultural technology and productivity. Some analytical considerations are also discussed in the context of vertical agricultural markets. Finally, an empirical investigation is carried out, revealing different implications of membership in the European Union (EU) and its predecessor, the European Economic Community (EEC), for productivity in the agricultural sector.

Background

Regional Trade Agreements (RTAs) represent a prominent feature of the current international trading system. For instance, the number of RTAs that have been reported to the World Trade Organization (WTO) exceeded 500 in 2012 (WTO, 2012). RTAs normally encompass reductions (or eliminations) of policy trade barriers (e.g., tariffs, non-tariff barriers) between member countries. Also, RTAs often bring about higher levels of business cooperation and trade facilitation measures (e.g., development of regional trade-enhancing infrastructure) between member countries that further promote intra-regional trade flows. RTAs are commonly associated with increases in intra-regional trade flows in agricultural products[a]. Naturally, such increases in intra-regional trade would be accompanied with a rise in market competition levels, in domestic agricultural markets for producers in importing member countries and in foreign regional agricultural markets for producers in exporting member countries.

The formation of an RTA is expected to generate various implications for agricultural technology and productivity in the importing and exporting member countries. This is because the formation of an RTA, which promotes intra-regional increases in agricultural trade flows, would eventually magnify market competition levels. In this context, several studies (e.g., Krugman, 1991; Frankel, 1997; Perdikis, 2007) noted that reductions in regional and bilateral trade barriers would not only decrease market prices, but would also compel firms in RTAs' member countries to upgrade their production technologies and to realize more efficient uses of inputs.

The RTA-induced intensification of market competition, realized through increases in import levels, could potentially provoke domestic firms in the agricultural sector to respond by upgrading their production technology and productivity levels to maintain or expand their domestic market shares vis-à-vis foreign regional exporting competitors. These potential responses are in accordance with many conventional studies indicating that increases in import competition levels would potentially induce domestic firms to upgrade their innovation activities (e.g., Pugel, 1978; Caves, 1985; Levinsohn, 1991). Furthermore, following the formation of a regional trading bloc, domestic firms in the agricultural sector become more exposed to foreign production technologies and practices. In this context, Josling (2011) noted that the formation of RTAs would promote flows of knowledge and technology between member countries, particularly from highly competitive agricultural sectors in some member countries to agricultural sectors in other member countries. This exposure to foreign knowledge and technology could generate supplementary incentives for domestic firms in the agricultural sector to invest in upgrading their production technologies and practices and realize the spillover effects.

Alternatively, the formation of an RTA could lessen the incentives of domestic firms in the agricultural sector of importing member countries to implement technology and productivity upgrading policies. These implications could arguably occur since RTA-induced increases in imports would raise the market competition levels and, consequently, would lead to reductions in the price-to-cost margins. This situation could reduce the market share of domestic firms and the expected returns from technology and productivity upgrading investments, and from innovation activities in general (Aghion and Howitt, 1998; Funk, 2003; Ghazalian, 2012).

RTA-induced increases in exporting activities would expose exporting firms in the agricultural sector of member countries to a more intense market competition in foreign regional agricultural markets, but also to new production technologies and practices. These exposures could potentially stimulate these exporting firms to upgrade their production technologies and productivity levels to further enhance their competitiveness in foreign regional markets and to realize the spillover effects. This argument is consistent with the seminal study on endogenous growth theory of Grossman and Helpman (1991) which identifies the positive effects of increases in exporting activities on innovation activities. Finally, the formation of a regional trading bloc generates larger effective agricultural markets for the exporting agricultural sectors of member countries. These larger agricultural markets could improve the feasibility of investments to upgrade agricultural technology and productivity. Lileeva and Trefler (2010) initially identified equivalent potential effects where larger effective regional markets would enhance the feasibility of investments in innovation activities.

The analytical literature has paid little attention to adjustments in agricultural technology and productivity, which are typically associated with shifts in supply curves, when examining the effects of RTAs on agricultural trade and welfare through the conventional frameworks. This paper contributes to the literature by analyzing the implications of RTA-induced adjustments in agricultural technology and productivity for the effects of RTAs on agricultural trade and welfare. It investigates the relationship between RTAs and agricultural technology and productivity in the context of Viner's (1950) conventional partial equilibrium framework with perfectly elastic foreign supply schedules faced by the importing member country. It also examines this relationship in the context of Pomfret's (1986) extended partial equilibrium framework depicting upward-sloping foreign supply schedules for the importing member country.

This paper shows changes in the initial benefits and losses from RTAs and depicts their redistributions following the RTA-induced adjustments in agricultural technology and productivity in the importing and exporting member countries. It is important to underscore the relevant study by Anania and McCalla (1995) that examined the implications of some domestic and trade policies, which are commonly used in developing countries, for the distribution of benefits from agricultural technology improvements. Using a partial equilibrium framework, Anania and McCalla (1995) identified different cases where policy interventions increase, reduce, or do not impact the overall benefits from agricultural technology improvements. Yet, in all cases, the distribution of benefits from agricultural technology improvements is found to be different with policy interventions.

A brief review of relevant literature

Many studies examined the feedback effects from international trade to productivity levels and innovation activities of firms and industries (e.g., Funk, 2003; Salomon and Shaver, 2005; Bloom et al., 2011; Ghazalian, 2012)[b]. Other studies focused on investigating the implications of improvements in regional and international market access for the productivity, technology, and innovation activities of firms and industries (e.g., Pavcnik, 2002; Baldwin and Gu, 2004; Costantini and Melitz, 2008; Ederington and McCalman, 2008; Lileeva and Trefler, 2010; Bustos, 2011).

Pavcnik (2002) examined the effects of trade liberalization, expressed through reductions in import policy barriers, on the productivity levels of domestic manufacturing plants in Chile. Pavcnik (2002) underscored rising trends in plant productivity levels in response to increases in market competition that is caused by the surge in import levels. These responses were found to be particularly prominent in the case of import-competing manufacturing sectors. Baldwin and Gu (2004) analyzed the responses of Canadian manufacturing plants to the continuous reductions in trade barriers between Canada and the rest of the world over time. They found that the continuous reductions in trade barriers have stimulated more Canadian manufacturing plants to participate in exporting activities. They also detected positive feedback effects from increases in exporting activities to productivity levels of individual plants[c].

Costantini and Melitz (2008) developed a dynamic model that explains the adjustments of firms to trade liberalization. Their model reveals that the anticipation of trade liberalization and the gradual implementation of trade liberalization schemes would provoke firms to innovate. Also, Ederington and McCalman (2008) used a dynamic model to show that international trade positively impacts firms' productivity levels and that the implementation of trade liberalization schemes would accelerate the rate of adoption (and diffusion) of new technology.

Lileeva and Trefler (2010) analyzed the implications of the Canada-United States Free Trade Agreement (CUSFTA) for the innovation activities and productivity levels of Canadian manufacturing plants. Specifically, they examined how Canadian manufacturing plants responded to reductions in United States' tariff rates. Lileeva and Trefler (2010) reported that Canadian manufacturing plants with lower productivity levels, which responded to the CUSFTA by starting to export or by exporting more, have realized important productivity gains. These plants have also experienced an increase in their domestic sales. Lileeva and Trefler (2010) showed that these plants have accelerated their adoption rates of new technologies and have increased their innovation rates of new products. Another recent study by Bustos (2011) examined the implications of the formation of the *Mercado Común del Sur* (MERCOSUR) or the Common Market of the South for the technological upgrading patterns of manufacturing firms in Argentina. The empirical results reveal that manufacturing firms in Argentina that benefited from relatively larger tariff reductions on their exports to Brazil have promoted their investments in upgrading production technology.

Methods

The seminal partial equilibrium framework of Viner (1950) indentifies the principal implications of RTAs through the trade creation and the trade diversion effects. The trade creation effect implies that the removal (or reduction) of policy trade barriers (e.g., tariffs) between RTA member countries leads to an increase in international trade flows between them, displacing less efficient domestic production. This effect generates positive welfare implications for the importing RTA member country. The trade diversion effect implies that the regional trade preferences induce a diversion of trade from the more efficient RTA non-member country to the less efficient RTA member country. This effect results in negative welfare implications for the importing RTA member country.

The benchmark analytical framework of Viner (1950) is illustrated through Figure 1. Consider one country "H" and two potential trade partners, country "F" and country "J". Country H is assumed to be small in economic sense, being unable to influence the international prices of the product under consideration. Hence, country H is assumed to be facing perfectly elastic supply schedules of foreign exporters. Let D_H and S_H represent the demand schedule and supply schedule of country H, respectively. Also, let S_F and S_J represent the perfectly elastic supply schedules of country F and country J, respectively[d]. Country J is assumed to be more efficient than country F with S_J placed below S_F. Initially, country H imposes a non-discriminatory tariff rate, denoted by τ, on imports coming from country F and from country J. Hence, the tariff-inclusive market prices of imported goods from country F and from country J are represented through $P_1 \equiv S_F(\tau)$ and $P_2 \equiv S_J(\tau)$, respectively. In this initial setup, country H ends up producing the quantity $[OQ_1]$ and importing the quantity $[Q_1Q_2]$ from country J. Country H does not import any quantity from country F.

The formation of an RTA that includes country H and country F as member countries dictates the removal of country H's tariff barriers on imports coming from country F, but keeps them imposed on those coming from country J. These preferences create trade flows from country F to country H, displacing less efficient country H's domestic

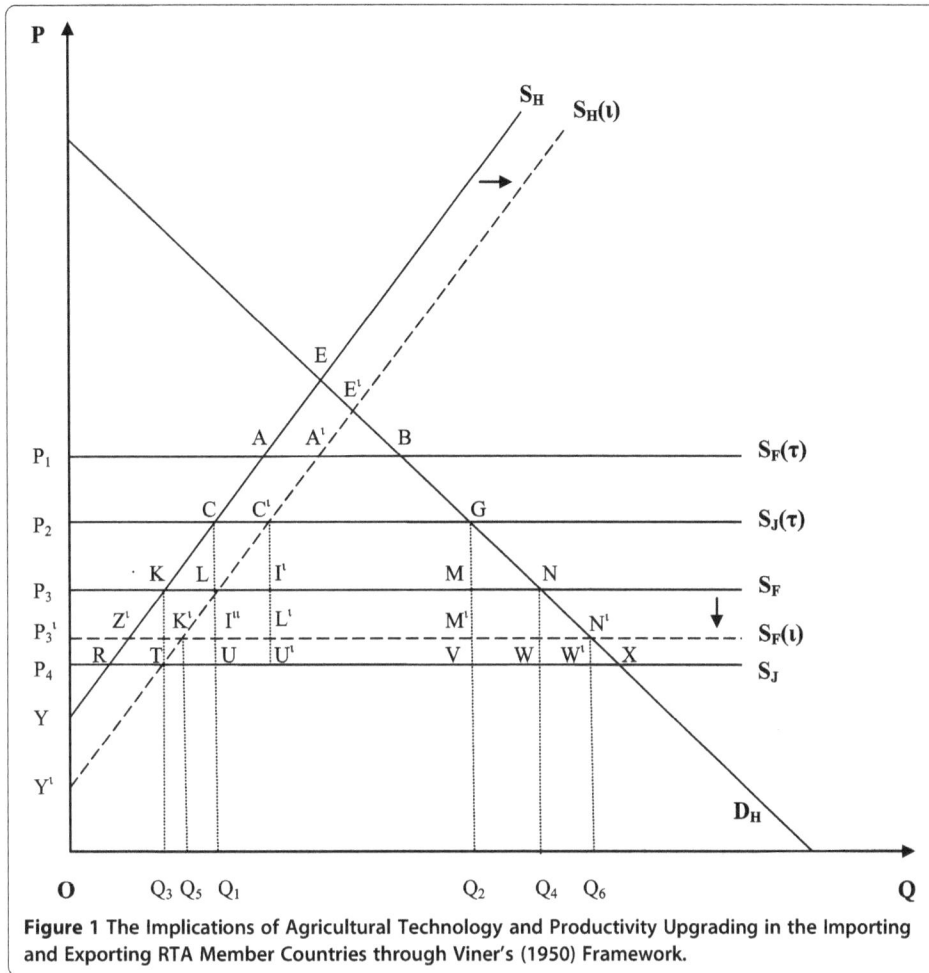

Figure 1 The Implications of Agricultural Technology and Productivity Upgrading in the Importing and Exporting RTA Member Countries through Viner's (1950) Framework.

production. However, it diverts trade from the more efficient country J to the less efficient country F. This is because the market price of imported goods from country F, which is now equivalent to $P_3 \equiv S_F$ is lower than the market price of imported goods from country J, which remains at $P_2 \equiv S_J(\tau)$. Consequently, country H now produces the quantity $[OQ_3]$ and imports the quantity $[Q_3Q_4]$ from the RTA member country F. The welfare analysis reveals that the RTA formation induces an increase in consumer surplus by the area $[P_2GNP_3]$. However, there is a decrease in producer surplus by the area $[P_2CKP_3]$. Also, tariff revenues, which were initially collected by country H's government from imports coming from country J (i.e., the quantity $[Q_1Q_2]$), are now lost. The loss in governmental tariff revenues is equivalent to the area $[CGVU]$. Summing up, the net effect on national welfare of country H from the RTA formation is equivalent to the area $[(CKL + GMN)-LMVU]^e$.

Next, following Pomfret (1986), we modify Viner's (1950) basic partial equilibrium framework to allow for upward-sloping supply schedules for both countries J and F. This is illustrated through Figure 2. In this case, both countries J and F could potentially end up exporting to country H, following the RTA formation. Let $IMPD_H$ represent the import demand schedule of country H, which is equal to the demand schedule minus the supply schedule of country H. Prior to the RTA formation, country H imports the quantity $[OQ_1]$ from country J and the quantity $[Q_1Q_2]$ from country F. The RTA formation induces a decrease in country H's imports from the non-member country J to $[OQ_3]$ and

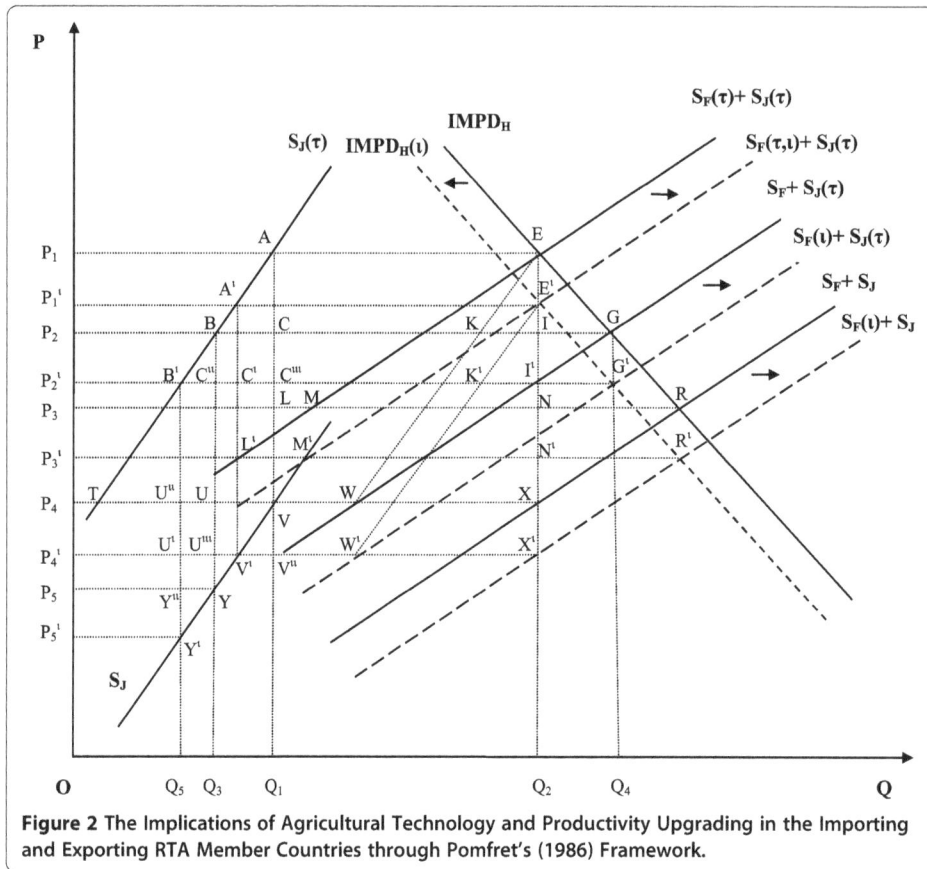

Figure 2 The Implications of Agricultural Technology and Productivity Upgrading in the Importing and Exporting RTA Member Countries through Pomfret's (1986) Framework.

an increase in country H's imports from the member country F to $[Q_3Q_4]$. Thus, the net welfare effect for the importing country H from the RTA formation is equivalent to the area $[(ABC + EIG)\text{-}BIXU + P_4UYP_5]$, where $[BIXU]$ depicts the losses from a higher expenditure on the quantity $[Q_3Q_2]$ that is imported from the RTA member country F whereas $[P4UYP_5]$ represents the gains from a lower expenditure on the quantity $[OQ_3]$ that remains imported from country J. Country F reaps benefits through an increase in welfare. These benefits are represented through the area $[CIXV]$, which depicts the gains from the post-RTA higher price received on the exported quantity $[Q_1Q_2]$, and through the area $[KGW]$, which depicts the increases in country F's producer surplus. Meanwhile, country J loses from the RTA formation. These losses are represented through the area $[P_4UYP_5]$, which depicts the effect of the post-RTA lower price received on the exported quantity $[OQ_3]$, and through the area $[UVY]$, which represents the remaining decreases in country J's producer surplus. Summing up, the net effect of the RTA formation on the global welfare is equivalent to the area $[(ABC + EIG + KGW)\text{-}BCVY]$.

Results and Discussion

The case of technology and productivity upgrading in the exporting and importing RTA member countries

RTAs are commonly associated with increases in intra-regional trade flows, causing a rise in market competition levels in domestic markets for importing countries and in foreign markets for exporting countries. As discussed earlier, the RTA-induced intensification of market competition in the importing country could potentially stimulate domestic agricultural producers to invest more in cost-reducing production technologies and to upgrade their productivity levels. Furthermore, domestic agricultural producers in the importing country could incorporate the RTA-induced flow of knowledge into their production routines and, hence, could realize the spillover effects from exposure to foreign technology and production practices. This scenario can be related to the studies of Pugel (1978), Caves (1985), and Pavcnik (2002), among others. Exporting agricultural producers, which encounter higher levels of market competition in regional markets following the RTA formation, are also compelled to upgrade their agricultural technology and productivity levels. They can similarly incorporate the knowledge spillover and learning-by-exporting effects into their production practices. This scenario can be related to the studies of Baldwin and Gu (2004), Lileeva and Trefler (2010), and Bustos (2011), among others.

The implications of these RTA-induced upgrading of agricultural technology and productivity levels are illustrated through Figure 1 using Viner's (1950) framework and through Figure 2 using Pomfret's (1986) framework. For country H, these upgrading adjustments are graphically depicted through a rightward parallel shift from S_H to $S_H(\iota)$ in Figure 1 and through a leftward parallel shift from $IMPD_H$ to $IMPD_H(\iota)$ in Figure 2. For country F, they are graphically expressed through a downward parallel shift and a rightward parallel shift from S_F to $S_F(\iota)$, as depicted in Figure 1 and Figure 2, respectively.

Figure 1 shows that country H's imports would increase from the pre-RTA quantity $[Q_1Q_2]$ to the post-RTA pre-adjustments quantity $[Q_3Q_4]$, which eventually increases further to the post-RTA post-adjustments quantity $[Q_5Q_6]^f$. Prior to the occurrence of any adjustment, the RTA formation induces an initial increase in the consumer surplus in country H by the area $[P_2GNP_3]$. Then, the RTA-induced downward shift from S_F to $S_F(\iota)$ would further augment the consumer surplus in country H by the additional area $[P_3NN'P_3^\iota]$. Also, prior to the occurrence of any adjustment, the RTA formation results in an initial decrease in the surplus of domestic agricultural producers in country H by the area $[P_2CKP_3]$. The occurrence of agricultural technology and productivity upgrading adjustments in country H would generate an increase in producer surplus by the area $[KLY'Y]$. Meanwhile, the occurrence of agricultural technology and productivity upgrading adjustments in country F would result in a decrease in producer surplus by the area $[P_3LK'P_3^\iota]$. Hence, following these adjustments, the net effect of the RTA formation on country H's producer surplus becomes equivalent to the area $[Z'K'Y'Y-P_2CZ'P_3^\iota]$. Accordingly, there is a higher likelihood of net gains in country H's producer surplus when the magnitude of the rightward shift from S_H to $S_H(\iota)$ is relatively larger than the magnitude of the downward shift from S_F to $S_F(\iota)$. Lastly, prior to the occurrence of these adjustments, the initial losses in governmental tariff revenues are equivalent to the basic area $[CGVU]$. Following the occurrence of these adjustments, the current implicit losses in governmental tariff revenues become equivalent to the smaller area $[C'GVU^\iota]^g$.

Thus, following the occurrence of these agricultural technology and productivity upgrading adjustments, the net welfare effect of the RTA formation for country H becomes equivalent to the area $[C'K'L^\iota + GM^\iota N^\iota + CC'Y'Y-L^\iota M^\iota VU^\iota]$. This net welfare effect can be readily compared to the initial net welfare effect (i.e., prior to the occurrence of these upgrading responses), which is equivalent to the area [CLK + GMN-LMVU]. Hence, these upgrading adjustments would raise the initial positive welfare effect of the RTA formation, expressed through the trade creation effect, by the supplementary area [LI'L'K' + MNN'M' + CC'Y'Y]. Concurrently, they would reduce the initial negative welfare effect of the RTA formation, expressed through the trade diversion effect, by the area [LI'U'U + I'MM'L']. Consequently, these upgrading adjustments would increase the probability to realize higher net positive welfare effects from the RTA formation[h].

Figure 2 presents the implications of the RTA-induced upgrading of agricultural technology and productivity levels using Pomfret's (1986) framework. For ease of exposition, Figure 2 illustrates the case where the shift margins between the parallel schedules $S_F(\tau) + S_J(\tau)$, $S_F + S_J(\tau)$, and $S_F + S_J$ are equivalent to those between the parallel schedules $S_F(\tau,\iota) + S_J(\tau)$, $S_F(\iota) + S_J(\tau)$, and $S_F(\iota) + S_J$. Also, following the RTA formation, Figure 2 illustrates the case where the pre-upgrading total imported quantity by country H is equivalent to the post-upgrading total imported quantity, remaining unchanged at $[OQ_4]$. Accordingly, following the RTA formation, the occurrence of agricultural technology and productivity upgrading adjustments causes increases in country F's exports to country H from the quantity $[Q_3Q_4]$ to the quantity $[Q_5Q_4]$ and decreases in country J's exports to country H from the quantity $[OQ_3]$ to the quantity $[OQ_5]$.

Following the occurrence of these upgrading adjustments, the net welfare effect for the importing country H changes from the area $[(ABC + EIG)\text{-}BIXU + P_4UYP_5]$ (see previous section) to the area $[(A'B'C' + E'I'G')\text{-}B'I'X'U' + P_4^lU'Y'P_5^l]$. Hence, the post-upgrading adjustments resulted in larger negative area that is augmented by [B'C''U''U'], reflecting a higher expenditure on the additional imports from country F. Also, the positive area is reduced by [U''UYY'']. The latter reflects the implications of the post-upgrading adjustments, leading to a smaller imported quantity from country J. In other words, following the RTA formation, the benefits from a lower expenditure on imports from country J at the post-upgrading adjustments equilibrium are smaller than those at the pre-upgrading adjustments equilibrium. Summing up, the occurrence of agricultural technology and productivity upgrading adjustments has resulted in a decrease in the benchmark net welfare effect of RTA formation for the importing country H.

Turning to the RTA exporting member country F, the post-upgrading gains from the RTA formation, which are equivalent to the area [C'I'X'V' + K'G'W'], are higher than the pre-upgrading gains, which are equivalent to the area [CIXV + KGW] (see previous section). Specifically, the net post-upgrading increases in country F's welfare, resulting from a higher price received on additional exports, is equivalent to the area [C'C''V''V']. Meanwhile, the pre-upgrading gains and the post-upgrading gains in country F's producer surplus (i.e., the areas [KGW] and [K'G'W'], respectively) are equivalent by construction under the current scenario. Turning to the non-RTA member country J, the losses from the RTA formation are attenuated, being reduced from the pre-upgrading adjustments area $[P_4UYP_5 + UVY]$ (see previous section) to the post-upgrading adjustments area $[P_4^lU'Y'P_5^l + U'V'Y']$. Hence, given that [UVY] is equivalent to [U'V'Y'] by construction, the post-upgrading losses are lower than the pre-upgrading losses by the area [U''UYY'']. The latter is derived from the lower quantities subjected to an RTA-induced lower price.

Finally, the pre-upgrading and the post-upgrading changes in global welfare remain equivalent through the scenario presented in Figure 2. In other words, we get:

$$[\Omega^{\iota}(\text{with RTA}) - \Omega^{\iota}(\text{without RTA})] = [\Omega(\text{with RTA}) - \Omega(\text{without RTA})] \tag{1}$$

where Ω^{ι} and Ω depict the global welfare with and without agricultural technology and productivity upgrading adjustments, respectively. Rearranging, Equation (1) can be expressed as:

$$[\Omega^{\iota}(\text{with RTA}) - \Omega(\text{with RTA})] = [\Omega^{\iota}(\text{without RTA}) - \Omega(\text{without RTA})] \tag{2}$$

Equation (2) implies that the changes in global welfare resulting from agricultural technology and productivity upgrading following the formation of RTA are equivalent to those that would result from a comparable upgrading that occurs in the absence of RTA. These results are reminiscent of those discussed in Anania and McCalla (1995), indicating that the global welfare gains from the adoption of new technology in the presence of distorting policies (e.g., export tax) are equivalent to the global welfare gains from the adoption of new technology under free trade.

The case of technology and productivity upgrading in the exporting RTA member country and downgrading in the importing RTA member country

Through this alternative scenario, the RTA formation induces the agricultural sector of the exporting member country F to upgrade its technology and productivity. However, the RTA formation forces the domestic agricultural sector in the importing member country H to reduce its budget allocated for technology and productivity upgrading and maintenance. Such reductions would eventually lead to decreases in productivity and efficiency levels. This scenario can be related to the studies of Aghion and Howitt (1998), Funk (2003), and Ghazalian (2012), among others. Figure 3 illustrates these adjustments through Viner's (1950) framework, where the supply curve of country F exhibits a downward parallel shift from S_F to $S_F(\iota)$ and where the supply curve of country H experiences a leftward parallel shift from S_H to $S_H(\iota)$. Hence, Figure 3 indicates that country H's imports would rise from the pre-RTA quantity $[Q_1Q_2]$ to the post-RTA pre-adjustments quantity $[Q_3Q_4]$, and further to the post-RTA post-adjustments quantity $[Q_5Q_6]$.

Figure 3 shows that the RTA formation would initially increase the consumer surplus by the area $[P_2GNP_3]$. This area is further increased by $[P_3NN'P_3^{\iota}]$ due to country F's agricultural technology and productivity upgrading responses. The producer surplus would initially decrease by the area $[P_2CKP_3]$ due to the increase in import competition levels brought about by the RTA formation. The producer surplus would experience a further reduction by the area $[P_3KZ^{\iota}P_3^{\iota}]$ due the RTA-induced downward shift of country F's supply curve and by the area $[K^{\iota}Z^{\iota}YY^{\iota}]$ due to the RTA-induced leftward shift of country H's supply curve. Also, following the RTA formation, the pre-adjustments losses in governmental tariff revenues are equivalent to the area $[CGVU]$. Comparatively, the post-adjustments losses in governmental tariff revenues become larger and are equivalent to the area $[C^{\iota}GVU^{\iota}]^{\iota}$.

Summing up, the post-adjustments net change in national welfare of country H is equivalent to the area $[KZ^{\iota}L^{\iota} + GM^{\iota}N^{\iota}-C^{\iota}CK-L^{\iota}M^{\iota}VU^{\iota}-K^{\iota}Z^{\iota}YY^{\iota}]$. This outcome can be compared to the pre-adjustments net change in national welfare of country H, which is

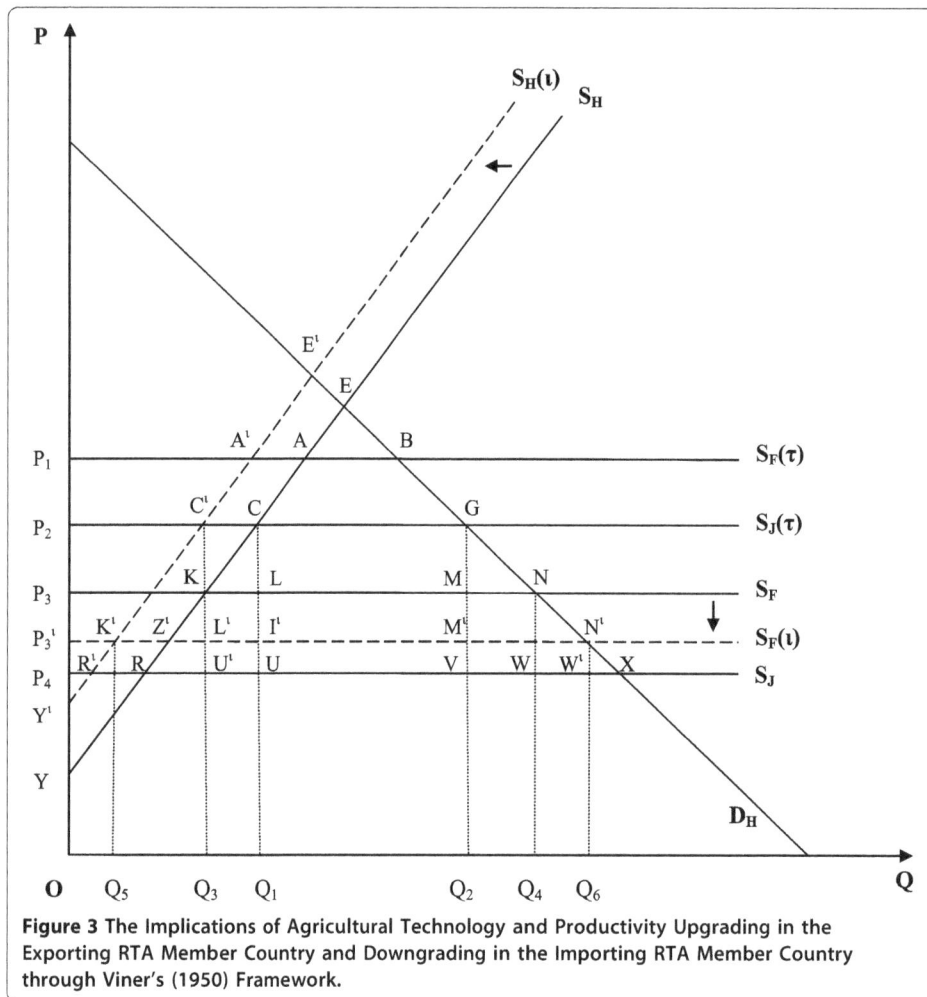

Figure 3 The Implications of Agricultural Technology and Productivity Upgrading in the Exporting RTA Member Country and Downgrading in the Importing RTA Member Country through Viner's (1950) Framework.

equivalent to the area [CKL + GMN-LMVU]. Accordingly, the adjustments in agricultural technology and productivity would result in an ambiguous change in the RTA's positive welfare effects, which are augmented by the area [KZ'L' + MNN'M'] and reduced by the area [CKL]. The likelihood of a post-adjustments increase in the RTA's positive welfare effects can occur with a larger downward shift from S_F to $S_F(\iota)$. These adjustments would also cause an ambiguous change in the RTA's negative welfare effects, which are magnified by the area [C'CK + L'I'UU' + K'Z'YY'], but are lessened by the area [LMM'I']^j. Hence, the larger is the downward shift from S_F to $S_F(\iota)$ and the smaller is the leftward shift from S_H to $S_H(\iota)$, the higher is the likelihood of a post-adjustments reduction in the RTA's negative welfare effects.

Figure 4 illustrates the corresponding scenario through Pomfret's (1986) framework. As in Figure 2, there are parallel rightward shifts in the export supply schedules with equivalent shift margins. Meanwhile, Figure 4 depicts a parallel rightward shift in the import demand curve. Also, following the RTA formation, Figure 4 illustrates the case where the corresponding pre-adjustments and post-adjustments equilibrium prices remain unchanged.

Following the RTA formation, Figure 4 indicates that the net welfare effect for the importing country H changes from the pre-adjustments area [(ABC + EIG)-BIXU + P_4UYP$_5$]

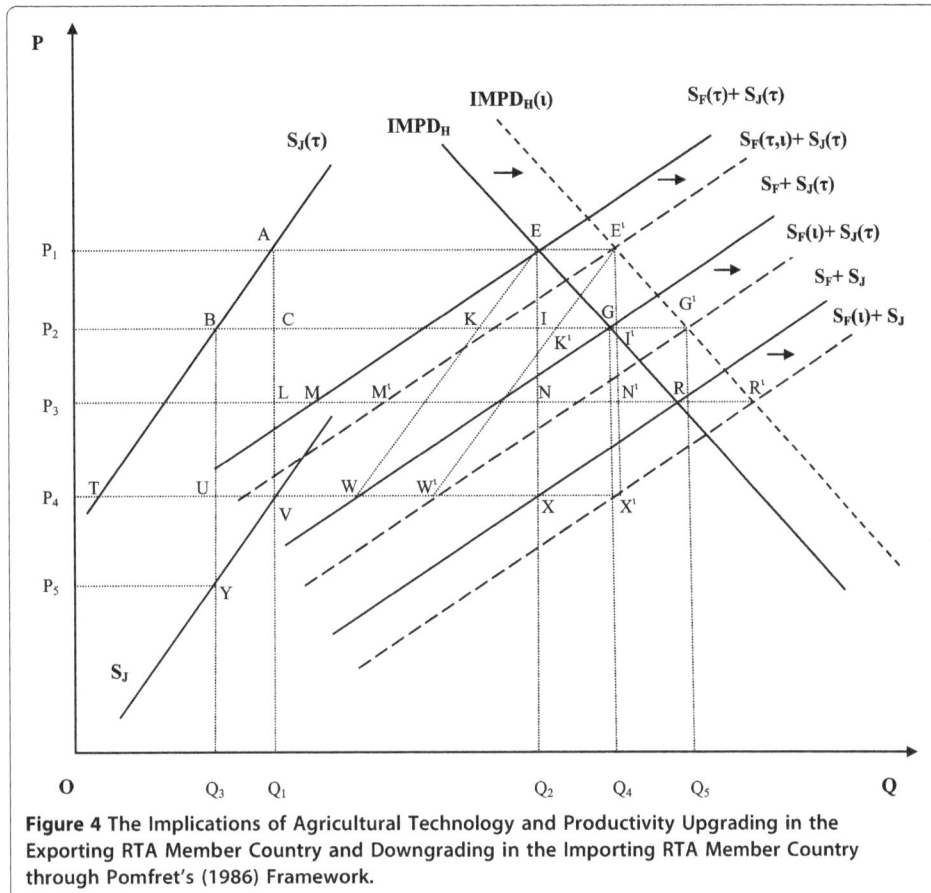

Figure 4 The Implications of Agricultural Technology and Productivity Upgrading in the Exporting RTA Member Country and Downgrading in the Importing RTA Member Country through Pomfret's (1986) Framework.

(see previous section) to the post-adjustments area $[(ABC + E^lI^lG^l)-BI^lX^lU + P_4UYP_5]$. Hence, there is a post-adjustments decrease in net welfare by the area $[II^lX^lX]$. These losses are associated with a higher expenditure on the quantity $[Q_2Q_4]$ that is imported from country F. Meanwhile, following the RTA formation, the gains of country F are increased from the pre-adjustments area $[CIXV + KGW]$ (see previous section) to the post-adjustments area $[CI^lX^lV + K^lG^lW^l]$. Hence, the post-adjustments increase in country F's gains is equivalent to the area $[IGX^lX]$. Finally, the pre-adjustments and post-adjustments losses of country J from the RTA formation remain unchanged at $[P_4UYP_5 + UVY]$ (see previous section). This outcome is resulting from the equivalent pre-adjustments and post-adjustments quantities that are imported by country H from country J (i.e., $[OQ_3]$) and from the equivalent pre-adjustments and post-adjustments equilibrium prices received by country J producers (i.e., P_5). The scenario presented in Figure 4 implies that the pre-adjustments and post-adjustments changes in global welfare are equivalent. Hence, the previously discussed Equation (1) and Equation (2) also pertain for this scenario.

Some analytical considerations in the case of vertical agricultural markets

RTAs commonly encompass preferential market access schemes covering vertically-related agricultural products. These regional trade preferences for imports of upstream (primary or intermediate) and downstream (final) products could have various effects on market competition levels. Accordingly, they could induce various interactive implications for technology and productivity of upstream and downstream agricultural industries along the supply chain.

Consider the scenario where country F is an exporter of the downstream product and an importer of the upstream product. Also, assume that the downstream agricultural industry in country F benefits from significant RTA-induced tariff reductions on imports of the upstream product that are purchased from country H. Consequently, the export supply schedule of country F's downstream product to country H would experience a downward/rightward shift caused by lower prices of the upstream product. This downward/rightward shift would increase exports and, hence, the regional market share of country F's downstream agricultural industry. Then, the previously discussed dynamic patterns could prevail once again where increases in market shares would further stimulate country F's downstream agricultural industry to upgrade its technology and productivity. These adjustments would result in a supplementary downward shift and rightward shift in $S_F(\iota)$ through Viner's (1950) framework (i.e., Figure 1 and Figure 3) and Pomfret's (1986) framework (i.e., Figure 2 and Figure 4), respectively.

Now, the RTA-induced growth in the exporting activities of country H's upstream agricultural industry would similarly enhance its technology and productivity. Consequently, this upstream agricultural industry would eventually supply its product at lower prices to the downstream agricultural industries in both RTA member countries H and F. Then, the downstream agricultural industry in country F would further increase its market share within the regional trading bloc, generating supplementary positive feedback effects on its technology and productivity.

The lower market prices of the upstream product could also enhance the competitiveness of country H's downstream agricultural industry. However, country H's downstream agricultural industry would be simultaneously facing higher import competition levels from country F. This is because, in this scenario, country F's downstream agricultural industry also benefits from the lower prices of the upstream product. For instance, consider the case where the increases in the downstream market competition levels, which are promoted by the lower prices of the upstream product, outweigh the corresponding competitiveness benefits for country H's downstream agricultural industry. Then, the previously discussed patterns could prevail where increases in import competition could either induce country H's downstream agricultural industry to upgrade its technology and productivity (as illustrated in the previous section) or to reduce its technology and productivity upgrading efforts (as illustrated in the previous section). Reversed responses could arguably prevail when the competitiveness benefits for country H's downstream agricultural industry, which are derived from the lower prices of the upstream product, more than compensate the corresponding effects of increases in the downstream market competition levels.

Next, consider an alternative scenario where country F is an exporter of the upstream and the downstream products. The RTA-induced increases in exports of country F's upstream and downstream agricultural industries would expand their regional market shares. These increases would stimulate technology and productivity upgrading activities in both industries. Thus, the downstream export supply schedule of country F would experience a basic downward shift in Viner's (1950) framework (i.e., Figure 1 and Figure 3) and a basic rightward shift in Pomfret's (1986) framework (i.e., Figure 2 and Figure 4) associated with the upgrading activities in the downstream industry. Also, the downstream export supply schedule of country F would experience a corresponding supplementary shift resulting from the lower prices of the upstream product brought about through the upgrading activities in the upstream industry.

The downstream agricultural industry in country H would be facing higher levels of import competition, which are derived from the technology and productivity upgrading efforts in country F's downstream and upstream industries. However, it would be simultaneously benefiting from the extra lower prices of the upstream product, which are derived from the upgrading efforts in country F's upstream industry. For instance, consider the case where the effects of the additional increases in import competition levels outweigh the benefits from the extra lower prices of the upstream product. Then, country H's downstream agricultural industry would respond by either upgrading its technology and productivity efforts (as illustrated in the previous section) or by reducing them (as illustrated in the previous section). Reversed responses could occur when the benefits from the extra lower prices of the upstream product outweigh the corresponding effects of increases in import competition levels.

Empirical investigation

This section presents an empirical investigation on the implications of membership in the European Union (EU) and its predecessor, the European Economic Community (EEC), for productivity in the agricultural sector. The productivity measure is derived from the World Bank Indicators' database, and it is represented through the value added per worker in the agricultural sector (data are presented in constant 2000 US dollars). The value added equals the value of output of the agricultural sector less the value of intermediate inputs. The agricultural sector is defined according to the International Standard Industrial Classification (ISIC), and it corresponds to the divisions 1–5. The dataset used through the empirical investigation covers observations for 14 European countries[k] over 29 years (1981–2009).

It is commonly perceived that the accession to the EU/EEC has led to increases in market competition levels faced by agricultural producers in member countries. This is because the EU/EEC has been Europe's (and world's) largest trading bloc and has been characterized by significant depth of economic integration and removal of trade barriers among member countries since formation. As discussed through this study, RTA-induced increases in market competition levels would likely cause adjustments in agricultural productivity. These adjustments are typically associated with shifts in supply curves. As shown through the analytical framework, shifts in supply curves would have implications for trade flows and national welfare. Given the dynamic nature of agricultural productivity patterns, the empirical specification compares productivity growth between EU/EEC member countries and other non-EU/EEC European countries representing the reference group. It is used to examine the occurrence of diverging shifts in supply curves of EU/EEC member countries from those of non-EU/EEC European countries. The estimating empirical model is a basic growth equation (Baumol, 1986; Caselli et al., 1996), and it is given by:

$$\ln(Y_{it}) - \ln(Y_{it-T}) = \beta \ln(Y_{it-T}) + \mathbf{EU_{it}}\boldsymbol{\delta} + \gamma_t + \varepsilon_{it} \tag{3}$$

where Y_{it} represents the productivity of country i at time t, $\mathbf{EU_{it}}$ is a row-vector of binary variables that take the value of one when the corresponding country i is a member of the EU/EEC at time t and zero otherwise. This vector includes a binary variable for the

Original Member Countries (OMCs)[l], binary variables for Denmark and the United Kingdom that accessed the EEC in 1973, binary variables for Portugal and Spain that become members of the EEC in 1986, and binary variables for Austria, Finland, and Sweden that become members of the EU in 1995. The reference group for these binary variables covers the European countries that are not members of the EU/EEC. It comprises countries that are continuing and previous members of the European Free Trade Association (EFTA) in our dataset[m]. The parameter γ_t is a time-specific effect, and ε_{it} is a stochastic error term. The regressions are implemented with five-year lags (i.e., $T = 5$).

The empirical results are presented in Table 1. Column (i) shows the estimated coefficients with robust standard errors. The estimated coefficient on the lagged productivity variable (i.e., $\hat{\beta}$) is negative and statistically significant at the 1% level. This finding indicates a convergence in productivity of the European countries in the agricultural sector over time. The implied convergence rate is 0.027 or 2.7%. The estimated coefficients on the EU/EEC binary variables for OMCs and for Denmark are positive and statistically significant at the 1% level. They indicate higher average productivity growth rates by 10.2%

Table 1 Empirical results

	(i)	(ii)	(iii)
Lagged Productivity	−0.125***	−0.130***	−0.122***
	(0.022)	(0.020)	(0.022)
EU/EEC member (OMCs)	0.097***	0.094***	0.087***
	(0.016)	(0.017)	(0.022)
EU/EEC member (Denmark)	0.113***	0.110***	0.103***
	(0.034)	(0.036)	(0.038)
EU/EEC member (United Kingdom)	−0.027	−0.029	−0.037
	(0.019)	(0.019)	(0.025)
EU/EEC member (Portugal)	−0.185***	−0.195***	−0.190***
	(0.045)	(0.044)	(0.047)
EU/EEC member (Spain)	0.040	0.034	0.032
	(0.032)	(0.032)	(0.034)
EU/EEC member (Austria)	−0.023	−0.029	−0.033
	(0.027)	(0.027)	(0.032)
EU/EEC member (Finland)	0.093***	0.088***	0.082**
	(0.029)	(0.029)	(0.034)
EU/EEC member (Sweden)	0.109***	0.106***	0.098***
	(0.025)	(0.024)	(0.032)
Trend		0.003***	
		(0.001)	
EFTA			−0.014
			(0.029)
Time-specific effects	Yes	No	Yes
Number of observations	336	336	336
R-squared	0.378	0.329	0.378

Notes: The dependent variable is productivity growth rate $\ln(Y_{it}) - \ln(Y_{it-T})$, where Y_{it} represents the productivity of country i at time t. Productivity is measured through the value added per worker (constant 2000 US dollars) in the agricultural sector. Robust standard errors are reported in parenthesis. The symbols "***", "**", and "*" denote statistical significance at the 1%, 5%, and 10% levels, respectively.

for OMCs and by 12.0% for Denmark compared to the average productivity growth rates of the European countries that are not members of the EU/EEC, *ceteris paribus*. Also, the estimated coefficients on the EU/EEC binary variables for Finland and Sweden, which joined the EU in 1995 and ceased to be member countries of the EFTA, are positive and statistically significant at the 1% level. They indicate higher average productivity growth rates by 9.7% and by 11.5% compared to the reference group, respectively. These results can be expressed through an RTA-induced shift in the supply curve (relative to the reference supply curve) as depicted through Figure 1 and Figure 2. This is consistent with an analytical scenario where the EU/EEC-induced intensification in market competition has provoked producers in the agricultural sector of OMCs, Denmark, Finland, and Sweden to respond by upgrading their production technology and productivity levels to maintain or expand their market shares vis-à-vis foreign regional competitors.

The estimated coefficient for Portugal's binary variable is found to be negative and statistically significant at the 1% level. It indicates that Portugal has lower productivity growth rates compared to the reference group by 16.9%, *ceteris paribus*. This result can be expressed through an RTA-induced shift in the supply curve (relative to the reference supply curve) as depicted through Figure 3 and Figure 4. This is consistent with an analytical scenario where the EU/EEC-induced raise in market competition levels has lessened the incentives of agricultural producers in Portugal to implement technology and productivity upgrading policies. This outcome could occur through decreases in price-to-cost margins, which reduce the market share of producers and the expected returns from technology and productivity upgrading investments. The estimated coefficients on the EU/EEC binary variables for the United Kingdom, Spain, and Austria are not statistically significant. Overall, the results suggest international differences in the implications of EU/EEC membership for productivity in the agricultural sector across the EU/EEC countries.

Column (ii) shows the results when substituting the time-specific effects with a time trend variable. The results remain equivalent to those presented in column (i). The estimated coefficient on the time trend variable is positive and statistically significant at the 1% level. Finally, we estimate the empirical model when including a binary variable that takes the value of one for the continuing member countries of EFTA (i.e., Iceland, Norway, and Switzerland) and zero otherwise. Hence, the reference group for the EU/EEC binary variables becomes the previous EFTA member countries (i.e., Austria, Finland, and Sweden) prior to their accession to the EU in 1995. The results, presented in column (iii), are similar to those reported in column (i).

Conclusions

This paper aims to enhance our understanding of the relationship between RTAs and agricultural technology and productivity. The implementation of an RTA is expected to induce increases in market competition levels, in domestic markets through increases in imports and in foreign markets through increases in exports. These increases in market competition levels could eventually result in important implications for agricultural technology and productivity in importing and exporting member countries of regional trading blocs. This paper examines these implications in the context of Viner's (1950) conventional partial equilibrium framework with perfectly elastic foreign supply schedules faced by the importing member country, and in the context of Pomfret's (1986) extended partial equilibrium framework depicting upward-sloping foreign supply schedules for the importing member country.

Different scenarios are implemented and discussed where RTA's initial benefits and losses experience considerable changes following the RTA-induced adjustments in agricultural technology and productivity in the importing and exporting member countries. The analysis underlines important redistributions of the benefits and losses between importing and exporting member countries and between consumers and producers. Also, the implications of these adjustments for global welfare are found not to be necessarily different in the presence or absence of regional trading blocs. This outcome is revealed through the illustrated scenarios which show equivalence of these implications with and without the RTA in place. This paper also presents and discusses some analytical considerations in the case of vertical agricultural markets where RTA preferential schemes cover imports of upstream (primary or intermediate) and downstream (final) products.

Following the analytical scenarios, this paper carries out an empirical investigation on the implications of membership in the EU/EEC for productivity in the agricultural sector. The results from a basic growth equation indicate that many EU/EEC member countries (OMCs, Denmark, Finland, and Sweden) have higher average productivity growth rates than non-EU/EEC European countries. These findings are consistent with the analytical scenarios where RTAs promote upgrading in productivity levels. Also, the results reveal that Portugal has lower productivity growth rates compared to non-EU /EEC European countries. This finding is consistent with the analytical scenarios where RTAs lessen the incentives to implement productivity upgrading policies. Overall, the results reveal international differences in the implications of the EU/EEC membership for productivity in the agricultural sector.

This paper provide analysts and policy-makers with an interactive analytical background accompanied with an empirical evidence on the relationship between RTAs and agricultural technology and productivity. This is essential when designing international trade, innovation, and productivity policies for the agricultural sector. Finally, this paper lends itself to a follow-up empirical study that estimates the feedback effects from the RTA-induced changes in agricultural technology and productivity to the trade creation and trade diversion effects of RTAs.

Endnote

[a]Many empirical studies (e.g., Sarker and Jayasinghe, 2007; Grant and Lambert, 2008; Lambert and McKoy, 2009; Sun and Reed, 2010; Ghazalian et al., 2011) reported increasing effects of several RTAs on international trade in agricultural products between member countries (i.e., trade creation effects), which often significantly exceed the decreasing effects of these RTAs on international trade in agricultural products between member countries and non-member countries (i.e., trade diversion effects).

[b]Funk (2003) found that increases in import competition levels have reduced Research and Development (R&D) activities of manufacturing firms in the United States. Meanwhile, increases in foreign sales, brought about by real exchange rate depreciation, have enhanced their R&D activities. Salomon and Shaver (2005) reported positive feedback effects from exporting activities to innovation activities. Bloom et al. (2011) found that European manufacturing firms have upgraded their productivity levels, and their

innovation and R&D activities in response to increases in imports from China. Ghazalian (2012) found that private R&D expenditures of the food processing sectors in the Organization for Economic Cooperation and Development (OECD) countries have responded positively to increases in exports but have reacted negatively to increases in import competition levels in domestic markets.

[c]Baldwin and Gu (2004) indentified several channels through which the growth of productivity levels occurred. These channels are: 1) exposure to higher levels of international competition in foreign markets, 2) learning-by-exporting where knowledge spillover and experience derived from exporting activities promote increases in industrial productivity levels, and 3) economies of scale realized through higher levels of product specialization.

[d]The supply schedules of country F and country J to country H encompass various international trade costs, particularly transportation, insurance, and information costs. For instance, consider the case where country H is geographically closer to country F than to country J. Thus, country F's exports require lower transportation and insurance costs than country J's exports. In this case, the wedge between S_F and S_J would reflect larger production efficiency differences.

[e]The likelihood that the welfare gains from the RTA formation (i.e., [CKL + GMN]) is larger than the welfare losses (i.e., [LMVU]) is function of several factors. For example, this likelihood increases when the wedge between S_F and S_J is relatively small. It also increases with a more elastic demand schedule of country H and with a higher non-discriminatory tariff rate imposed by country H.

[f]Alternatively, country H's total imports can decrease when the magnitude of the rightward shift from S_H to $S_H(\iota)$ is significantly larger than the magnitude of the downward shift from S_F to $S_F(\iota)$.

[g]Alternatively, when assuming that the hypothetical annulment of the RTA would induce a shift in country H's supply schedule back to its pre-RTA formation position (i.e., S_H), the implicit losses in governmental tariff revenues can be set at [CGVU].

[h]Domestic and foreign agricultural producers could be also compelled to upgrade their marketing strategies and to introduce new product attributes. Such adjustments would be graphically expressed through a rightward shift in the demand curve of country H (i.e., D_H), implying larger positive welfare effects.

[i]Alternatively, consider the case where the annulment of the RTA, which lessens the import competition level in the domestic market, induces a shift in country H's supply schedule back to its pre-RTA formation position (i.e., back to S_H). Then, the losses in governmental tariff revenues can be set at [CGVU].

[j]As discussed in endnote h, a rightward shift in the demand curve of country H (i.e., D_H) can also occur, leading to supplementary positive welfare effects.

[k]These countries are: Austria, Denmark, Finland, France, Germany, Iceland, Italy, the Netherlands, Norway, Portugal, Spain, Sweden, Switzerland, and the United Kingdom.

[l]The OMCs in our dataset are: France, Germany, Italy, and the Netherlands.

[m]The continuing members of the EFTA in our dataset are: Iceland, Norway, and Switzerland. Austria, Finland, and Sweden ceased to be EFTA members when they become members of the EU in 1995.

References

Aghion P, Howitt P (1998) Endogenous Growth Theory. Massachusetts Institute of Technology (MIT) Press, Cambridge, MA

Anania G, McCalla AF (1995) Assessing the Impact of Agricultural Technology Improvements in Developing Countries in the Presence of Policy Distortions. European Review of Agricultural Economics 22(1):5–24

Baldwin JR, Gu W (2004) Trade Liberalization: Export-market Participation, Productivity Growth and Innovation. Oxford Review of Economic Policy 20(3):372–392

Baumol WJ (1986) Productivity Growth, Convergence, and Welfare: What the Long-Run Data Show. Am Econ Rev 76(5):1072–1085

Bloom N, Draca M, van Reenen J (2011) "Trade Induced Technical Change: The Impact of Chinese Imports on Innovation, IT, and Productivity". In: Working Paper No. 16717. National Bureau of Economic Research (NBER), Cambridge, MA

Bustos P (2011) Trade Liberalization, Exports, and Technology Upgrading: Evidence on the Impact of MERCOSUR on Argentinian Firms. Am Econ Rev 101(1):304–340

Caselli F, Esquivel G, Lefort F (1996) Reopening the Convergence Debate: A New Look at Cross-Country Growth Empirics. Journal of Economic Growth 1(3):363–389

Caves RE (1985) International Trade and Industrial Organization: Problems, Solved and Unsolved. European Economic Review 28(3):377–395

Costantini J, Melitz MJ (2008) The Dynamics of Firm-Level Adjustment to Trade Liberalization. In: Helpman E, Marin D, Verdier T (eds) The Organization of Firms in a Global Economy. Cambridge, MA, Harvard University Press

Ederington J, McCalman P (2008) Endogenous Firm Heterogeneity and the Dynamics of Trade Liberalization. Journal of International Economics 74(2):422–440

Frankel JA (1997) Regional Trading Blocs in the World Economic System. Institute for International Economics, Washington, DC

Funk M (2003) The Effects of Trade on Research and Development. Open Economies Review 14(1):29–42

Ghazalian PL (2012) Assessing the Effects of International Trade on Private R&D Expenditures in the Food Processing Sector. Industry and Innovation 19(4):349–369

Ghazalian PL, Larue B, Gervais J-P (2011) Assessing the Implications of Regional Preferential Market Access for Meat Commodities. Agribusiness: An International Journal 27(3):292–310

Grant JH, Lambert DM (2008) Do Regional Trade Agreements Increase Members' Agricultural Trade? American Journal of Agricultural Economics 90(3):765–782

Grossman GM, Helpman E (1991) Innovation and Growth in the Global Economy. MIT Press, Cambridge, MA

Josling T (2011) Agriculture. In: Chauffour J-P, Maur J-C (eds) Preferential Trade Agreement Policies for Development: A Handbook. Washington, DC, The World Bank

Krugman P (1991) Is Bilateralism Bad? In: Helpman E, Razin A (eds) International Trade and Trade Policy. Cambridge, MA, MIT Press

Lambert D, McKoy S (2009) Trade Creation and Diversion Effects of Preferential Trade Associations on Agricultural and Food Trade. Journal of Agricultural Economics 60(1):17–39

Levinsohn J (1991) "Testing the Imports-as-Market-Discipline Hypothesis". Working Paper No. 3657. National Bureau of Economic Research (NBER), Cambridge, MA

Lileeva A, Trefler D (2010) Improved Access to Foreign Markets Raises Plant-Level Productivity… for Some Plants. The Quarterly Journal of Economics 125(3):1051–1099

Pavcnik N (2002) Trade Liberalization, Exit, and Productivity Improvements: Evidence from Chilean Plants. Rev Econ Stud 69(1):246–276

Perdikis N (2007) Trade Agreements: Depth of Integration. In: Kerr WA, Gaisford JD (eds) Handbook on International Trade Policy. Cheltenham, UK, Edward Elgar Publishing

Pomfret R (1986) The Theory of Preferential Trading Arrangements. Review of World Economics 122(3):439–465

Pugel TA (1978) International Market Linkages and U.S. Manufacturing. Ballinger Publishing, Cambridge, MA

Salomon RM, Shaver JM (2005) Learning by Exporting: New Insights from Examining Firm Innovation. Journal of Economics and Management Strategy 14(2):431–460

Sarker R, Jayasinghe S (2007) Regional Trade Agreements and Trade in Agri-food Products: Evidence for the European Union from Gravity Modeling Using Disaggregated Data. Agricultural Economics 37(1):93–104

Sun L, Reed MR (2010) Impact of Free Trade Agreements on Agricultural Trade Creation and Trade Diversion. American Journal of Agricultural Economics 92(5):1351–1363

Viner J (1950) The Customs Union Issue. Carnegie Endowment, New York, NY

World Trade Organization (WTO) (2012) Regional Trade Agreements. Retrieved November., from http://www.wto.org/english/tratop_e/region_e/region_e.htm

2

Information and output in agricultural markets: the role of market transparency

Christian Ahlers[1], Udo Broll[2]* and Bernhard Eckwert[3]

*Correspondence:
udo.broll@tu-dresden.de
[2]Department of Business and Economics, School of International Studies, Technische Universität Dresden, Helmholtzstr. 10, 01069 Dresden, Germany
Full list of author information is available at the end of the article

Abstract

This study is concerned with the impact of changes in market transparency on agricultural production levels. Market transparency is of central importance in the agri-food system as it affects the degree of uncertainty farmers face when taking economic decisions. In our study, we endogenize uncertainty by establishing a link between market transparency and the terms of contracting on the futures market. We find that a higher degree of market transparency leads to higher expected profits but does not increase agricultural production levels per se. However, when farmers have no access to futures markets, transparency does increase ex ante expected uility and output.

Keywords: Market transparency; Agriculture; Price risk; Information

Background

Markets for important food staples as grains and vegetable oils have seen a great deal of turmoil over the past decade. Particularly, uncertainty associated with price volatility negatively affects investment and production decisions by risk-averse farmers. An intuitive illustration is given by Figure 1 that displays the FAO real monthly food as well as the respective cereals price index. The data shows that global food markets exhibited serious patterns of uncertainty over the past decade.

However, the role of technology adoption and adjustment of supply is of central importance with respect to securing adequate levels of agricultural production and achieving global food security. Consequently, recent studies stressed the role of market transparency and information flows for improving production decisions in agricultural markets (see UNCTAD 2011). Thus, market transparency can be understood in terms of parameters that, at least partially, are under control of public and private agencies. Governmental and non-governmental organizations then may enhance transparency by providing a greater deal and more reliable information on important determinants of price movements, such as indicators of climate change, international market conditions, government price stabilization schemes, weather forecasts, disaster relief programs, stricter food safety standards, insurance and alternatives (UNCTAD 2011).

Against this background, the present study is concerned with the impact of changes in markets transparency on agricultural production levels. In particular, we study how the

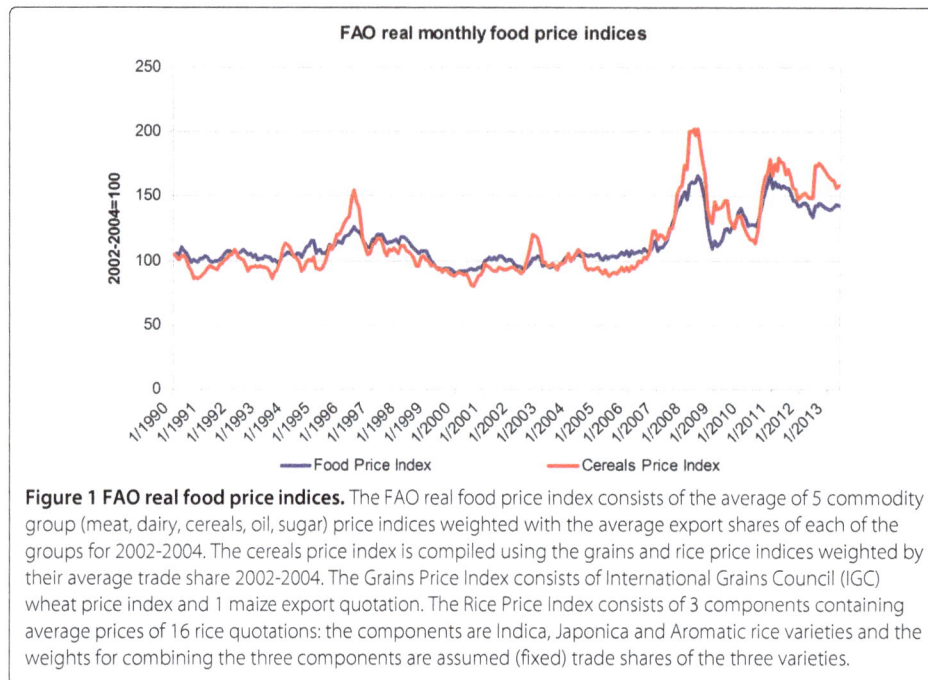

Figure 1 FAO real food price indices. The FAO real food price index consists of the average of 5 commodity group (meat, dairy, cereals, oil, sugar) price indices weighted with the average export shares of each of the groups for 2002-2004. The cereals price index is compiled using the grains and rice price indices weighted by their average trade share 2002-2004. The Grains Price Index consists of International Grains Council (IGC) wheat price index and 1 maize export quotation. The Rice Price Index consists of 3 components containing average prices of 16 rice quotations: the components are Indica, Japonica and Aromatic rice varieties and the weights for combining the three components are assumed (fixed) trade shares of the three varieties.

ex ante expected volume of agricultural production responds to additional information when information also affects the futures market for agricultural commodities. Earlier studies addressing this issue have modeled market transparency solely by means of exogenous changes in the distributional parameters of the market price. In such a framework, the futures market is completely separated from the underlying transparency concept (see Newbery and Stiglitz 1981, Kawai and Zilcha 1986, Frechette 2000, Moschini and Hennessy 2001, Allen and Lueck 2003 as well as Hudson 2007).

In our view, this approach misses out an important link that exists between market transparency and risk-sharing opportunities on futures markets. In our model, we endogenize uncertainty through an information system provided by governmental and nongovernmental organizations or related agencies that explicitly links market transparency to the terms of contracting on the futures market.

The notion of market transparency underlying this approach is adopted from the work of Eckwert and Zilcha (2001, 2003). Along the seminal contributions of Blackwell (1953) and Hirshleifer (1971), market transparency is linked to the informativeness of an observable signal which is (imperfectly) correlated with the future spot price. The uncertainty the farmer is exposed to then depends on the observed signal as well as on the information system within which the signal can be interpreted. We find that a higher degree of market transparency leads to higher expected profits but does not increase agricultural production levels per se. However, when we vary the prevailing risk-sharing regime such that farmers have no access to futures markets, transparency does increase ex ante expected utility and output.

In particular, we consider a model where a farmer faces risky revenues due to a random product price. The distribution of the price is given and there exists a futures market (see, for example, Broll et al. 2013). The terms at which the farmer can hedge the revenue risk through trade in futures contracts negatively depends on the degree of market transparency. Higher market transparency then affects the farmer's production decision in two opposing ways. Firstly, price uncertainty declines and ex ante expected utility increases, resulting in higher production levels. Secondly, additional information on future market conditions may interfere with the operation of risk-sharing markets and thereby decrease ex ante expected utility and production levels.

The paper is organized as follows. In section 'Methods', we introduce the model and our concepts of information and market transparency. Our main results are derived in section 'Results and discussion'. Section 'Conclusions' concludes.

Methods

The farmer produces in period 0 and sells her products in period 1 for a random price. Production costs, $c(q)$, are a strictly increasing and convex function of the volume of production q. The firm's random revenues, as of date 1, are $\tilde{p}q$, where $\tilde{p}$ is the one period ahead spot market price. The tilde refers to the random nature of the spot price which assumes values in $\Omega := [\underline{p}, \overline{p}]$, where $0 < \underline{p} < \overline{p} < \infty$.

Prior to choosing a production level, the farmer observes a signal s from a governmental or non-governmental organization and/or a related agency. This signal is the realization of a random variable $\tilde{s}$ which is correlated with $\tilde{p}$. The signal contains public information about the unknown future market price p. Thus, at the time when input and production decision are made, the relevant price expectation is the updated in a Bayesian way. Figure 2 depicts how the sequence of events unfolds in the model.

Futures markets for agricultural commodities open at date 0 after the signal has been observed. Let h be the future commitment of the farmer, i.e., h denotes the number of futures contracts sold by the farmer. To focus on the farmer's hedging motive as opposed to a speculative motive, we assume that the commodity derivative is unbiased. Therefore, the forward rate $f_0(s)$ is equal to the conditional mean of a contract's payoff,

$$f_0(s) = \mathrm{E}[\tilde{p}|s]. \tag{1}$$

Both, the payoff and the purchase price of the contract fall due in period 1.

The farmer's decision problem

The risk-averse farmer is maximizing expected utility, defined over random profits, $\tilde{\Pi}$, i.e.

$$\tilde{\Pi} = \tilde{p}q - c(q) - (\tilde{p} - f_0(s))h. \tag{2}$$

The decision making problem can be written as

$$\max_{q,h} \mathrm{E}[U(\tilde{\Pi})|s], \tag{3}$$

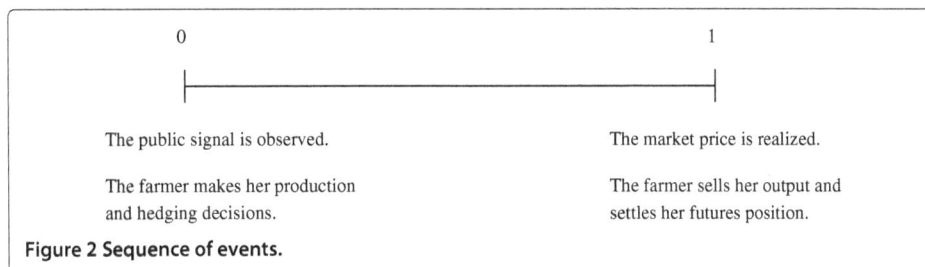

Figure 2 Sequence of events.

0   1

The public signal is observed.

The farmer makes her production and hedging decisions.

The market price is realized.

The farmer sells her output and settles her futures position.

where U is a strictly increasing, strictly concave and twice continuously differentiable utility function. The necessary first-order conditions, which are also sufficient, are

$$\mathrm{E}[\,U'(\tilde{\Pi})(\tilde{p} - c'(q))|s] = 0, \tag{4}$$

$$\mathrm{E}[\,U'(\tilde{\Pi})(f_0(s) - \tilde{p})|s] = 0, \tag{5}$$

where U' and c' are denoting marginal utility and marginal cost of production. From (4), (5) and the unbiasedness assumption, we obtain the optimal levels of production and hedging:

$$c'(q(s)) = f_0(s) \quad \text{and} \quad h = q(s). \tag{6}$$

For our model, equation (6) establishes the validity of the separation theorem and the full-hedging hypothesis. Thus, in the presence of risk-sharing markets, entrepreneurial decisions are independent of attitudes towards risk and, moreover, all risks will be fully hedged if the futures market is unbiased (see, for example, Wong 2007). Next we define our notion of market transparency. Following Eckwert and Zilcha (2001, 2003), transparency will be linked to the informational content of the signal s under different information systems and risk-sharing regimes.

Information systems and market transparency

We identify the transparency of a market with the informativeness of the signal $s \in S$, which is provided by a governmental or non-governmental organization or a related agency. The informativeness of a given signal then depends on the information system within which the signal can be interpreted. An information system, denoted by g, specifies for each state of nature, p, a conditional probability function over the set of signals: $g(s|p)$. The positive real number $g(s|p)$ defines the conditional probability (density) that the signal s will be observed if the true (yet unknown) state of nature is p_1. Using Bayes' rule, the farmer revises her expectations and maximizes utility on the basis of the updated beliefs.

The density for the prior distribution over S is given by

$$v(s) = \int_S g(s|p)\pi(p)\,\mathrm{d}p \quad \text{for all } s, \tag{7}$$

where π denotes the Lebesgue density function for the prior distribution over S. The density function for the updated posterior distribution over S is

$$v(p|s) = g(s|p)\pi(p)/v(s). \tag{8}$$

Blackwell (1953) suggested a criterion that ranks different information systems according to their informational content. Suppose g^1 and g^2 are two information systems with associated density functions $v^1(\cdot)$ and $v^2(\cdot)$. The following criterion induces an ordering on the set of information systems.

Information systems. The information system g^1 is said to be more informative than g^2 if there exists an integrable function λ such that

$$\int_S \lambda(s',s)\,ds' = 1, \tag{9}$$

holds for all s, and

$$g^2(s'|p) = \int_S g^1(s|p)\lambda(s',s)\,ds \tag{10}$$

holds for all $p \in S$.

According to this criterion $g^1 \succ_{\inf} g^2$, holds if g^2 can be obtained from g^1 through a process of randomization. The probability density $\lambda(s', s)$ in equation (10) transforms a signal s into a new signal s'. If the s'-values are generated in this way, the information system g^2 can be interpreted as being obtained from the information system g^1 by adding random noise. Therefore, the signals under information system g^2 convey no information about the value of $\tilde{p}$ that is not already conveyed by the signals under information system g^1. As a consequence, the a priori expected posterior price uncertainty under g^1 will be lower than under g^2.

As noted before, our notion of market transparency is based on the informational content of the signal s. We characterize a market as more transparent if the signal conveys more precise information about the future market price.

Market transparency. Agri-food markets are said to be more transparent under g^1 than under g^2 if $g^1 \succ_{\inf} g^2$.

The following lemma contains a property of information systems that turns out to be a very convenient tool for economic analysis. The lemma formulates a transparency criterion that is equivalent to the condition stated above.

Lemma 1. The agri-food market is more transparent under g^1 than under g^2 if and only if

$$\int_S F(v^1(\cdot|s)) v^1(s)\, ds \geq \int_S F(v^2(\cdot|s)) v^2(s)\, ds$$

holds for every convex function $F(\cdot)$ on the set of density functions over Ω.

A proof of Lemma 1 can be found in Kihlstrom (1984). Note that $v^1(\cdot|s)$ and $v^2(\cdot|s)$ are the posterior beliefs under the two information systems. Thus, Lemma 1 implies that higher transparency (weakly) raises the expectation of any convex function of posterior beliefs. In other words, additional information is equivalent to introducing a mean preserving spread in the probability of success. If the maximum expected utility is a convex function of the probability of success, information raises the ex ante expected utility of any risk-averse farmer.

Results and discussion

We now turn to the question how the ex ante expected volume of agricultural output is affected when the market becomes more transparent, i.e. when additional public information about future market conditions becomes available. We consider two cases. First, we assume that the farmer has full access to futures markets to hedge her revenue risk. In a second step, we consider a world where the farmer has no access to futures markets or other risk-sharing opportunities.

Impact of market transparency in the presence of futures market

The production, q, is contingent on the signal s. We define the volume of production, Q, as the average agricultural output before the signal has been observed,

$$Q = E_s[q(s)] = \int_S q(s) v(s)\, ds \tag{11}$$

Now we characterize the impact of higher market transparency on the volume of production in terms of the curvature of the marginal cost function.

Proposition 1. Higher transparency in the agricultural market leads to a higher (lower) expected level of supply if, and only if, the marginal cost function, $c'(q)$, is strictly concave (convex).

Proof of Proposition 1. It is evident from equation (1) that f_0 is a linear function of the posterior probability density function, $v(p|s)$. Lemma 1 and equation (11) imply that the average level of supply increases (decreases) with higher transparency in the agricultural market if, and only if, $q(f_0)$ is strictly convex (concave) in f_0. From equation (6) we obtain

$$q''(f_0) = -\frac{c'''[q(f_0)]}{c''[q(f_0)]^2}. \tag{12}$$

The desired result follows from equation (12).

As just seen, the impact of higher transparency on agricultural production depends only on the curvature of the marginal cost function. Concavity of the marginal cost function is a pattern sometimes seen in agriculture where the presence of fixed input factors (e.g. land and building of fixed size) constrains production.

By contrast, production processes that exhibit high substitutability among input factors may give rise to convex marginal cost functions. If the cost function is quadratic, i.e. the marginal cost function is linear, production will not be affected by changes in market transparency.

The result in Proposition 1 differs from the role attributed to market transparency in earlier studies which have modeled changes in transparency simply by means of an exogenous change in the distributional parameters of the market price. Our approach implies that market transparency affects the terms of contracting on the futures market and, hence, the farmer's production decision and output. The endogenous terms of futures contracting, therefore, constitutes an important link through which market transparency may affect the volume of supply in agricultural markets.

To illustrate our result and the impact of market transparency on the expected level of agricultural supply, we now consider an example with binary information structure. Under information system g^1, the signal provides full information about the future market price. Under information system g^2 the signal is completely uninformative.

Example
The farmer's cost function takes the form $c(q) = (1/a)q^a$, $a > 1$. The signal space contains two signals, s_1 and s_2. The market price takes the values p_1 and p_2, $p_1 < p_2$ with prior probabilities equal to $1/2$. The information system is given by the conditional probabilities $g(s_1|p_1) = g(s_2|p_2) = \beta$, and $g(s_1|p_2) = g(s_2|p_1) = 1 - \beta$, $\beta \in [1/2, 1]$. The parameter β measures the transparency of the agricultural market: for $\beta = 1$ the information system is fully informative indicating a transparent market; for $\beta = 1/2$ the information system is fully uninformative indicating an intransparent agricultural market.

The prior signal probabilities are

$$n(s_1) = g(s_1|p_1)f(p_1) + g(s_1|p_2)f(p_2) = n(s_2) = 1/2. \tag{13}$$

The posterior state probabilities can be calculated as

$$v(p_1|s_1) = v(p_2|s_2) = \beta \quad \text{and} \quad v(p_2|s_1) = v(p_1|s_2) = 1 - \beta. \tag{14}$$

From (13) and (14) we get

$$f_0(s_1) = p_1 + (p_2 - p_1)(1 - \beta), \tag{15}$$

$$f_0(s_2) = p_1 + (p_2 - p_1)\beta. \tag{16}$$

The separation theorem (see equation (6)) implies $q = (f_0)^{1/(a-1)}$. Therefore, ex ante expected output, Q, reads

$$Q = \frac{1}{2} \left\{ [p_1 + (p_2 - p_1)(1 - \beta)]^{1/(a-1)} + [p_1 + (p_2 - p_1)\beta]^{1/(a-1)} \right\}. \tag{17}$$

Now we compare the expected output supply in a transparent market, Q_t, with the expected output volume in an intransparent agricultural market, Q_{it}. Setting $\beta = 1$ and $\beta = 1/2$ we get

$$Q_t = \frac{1}{2} [p_1^{1/(a-1)} + p_1^{1/(a-1)}], \tag{18}$$

$$Q_{it} = \frac{(p_1 + p_2)^{1/(a-1)}}{2}. \tag{19}$$

Let us choose $p_1 = 5, p_2 = 15$ and $a = 11/10$. Then, the ratio Q_t/Q_{it} is much larger than 1. If a takes values larger than 2, then ratio Q_t/Q_{it} is smaller than 1. In that case, higher market transparency reduces the expected volume of agricultural supply. However, the effect tends to be small.

Before we vary the prevailing risk-sharing regime, we now turn to the impact of changes in market transparency on the farmer's expected profits.

Proposition 2. Higher transparency in the agricultural market always leads to an increase in the farmer's expected profits. Ex ante expected profits

$$E(\tilde{\Pi}) = \int_S \Pi(s) v(s) \, ds \tag{20}$$

are higher under g^1 than under g^2.

Proof of Proposition 2. Proceeding along the same lines as in the proof of Proposition 1 we need to show that

$$\Pi(f_0) = f_0 q(f_0) - c(q(f_0)) \tag{21}$$

is a convex function of $f_0(s)$. In the optimum, differentiating with respect to f_0 yields

$$\Pi'(f_0) = q(f_0) > 0 \quad \text{and} \quad \Pi''(f_0) = q'(f_0) > 0. \tag{22}$$

An increase in f_0 has a first-order effect on the farmer's maximum profit through the revenues $f_0 q(f_0)$. Since the farmer sells more when f_0 increases, this first-order effect on $\Pi(f_0)$ is stronger for larger values of f_0 and weaker for lower values. As a result, the farmer's profit function is convex in the forward rate. A more transparent agricultural market makes f_0 more sensitive to changes in the public signal, thereby leading to higher expected profits.

According to Proposition 2, higher market transparency leads to higher expected profits regardless of attitudes towards risk and of technological parameters as long as the cost function is convex. This result does not imply, however, that the farmer will be better off in terms of ex ante expected utility. When the signal affects an insurable risk, like in our model, the value of additional information depends on two opposing effects.

Firstly, when the farmer receives more reliable information she is able to improve her decisions, thereby increasing ex ante expected utility (Blackwell effect). Secondly, as was pointed out by Hirshleifer (1971), additonal information may interfere with the operation of risk-sharing markets thereby destroying some risk-sharing opportunities. Since the farmer is risk-averse, ex ante expected utility declines. Due to these opposing effects, the overall impact of higher market transparency on the farmer's ex ante expected utility is ambiguous.

Impact of market transparency in the absence of futures market

To further illustrate the interaction between the Blackwell and the Hirshleifer effect, we now consider the case when the farmer has no access to futures markets or other risk-sharing opportunities. The above results suggest that the (negative) Hirshleifer effect vanishes and the ex ante expected utility depends on the (positive) Blackwell effect only.

Proposition 3. If agricultural commodities cannot be traded on a futures market at date 0, ex ante expected utility is higher under information system g^1 than under g^2.

Proof of Proposition 3. In the absence of a futures market, the first-order condition for a farmer's decision problem reads

$$\mathrm{E}[\,U'(\tilde{\Pi})(\tilde{p} - c'(q))|s] = 0. \tag{23}$$

Denote the unique solution to (23) by $q(s)$ and define

$$U(q(s), p) = U[\,pq(s) - c(q(s))]\,.$$

The value function is

$$V(v(\cdot)|s) = \int_S U(q(s), p) v(p|s) dp. \tag{24}$$

To proof our proposition, we have to show that the value function is convex in the posterior belief $v(\cdot|s)$. Assume $v(\cdot|s) = \beta v^1(\cdot|s) + (1-\beta)v^2(\cdot|s), \beta \in [0, 1]$. Denote by $q^1(s)$ and $q^2(s)$ the optimal agricultural supply under the posterior beliefs $v^1(\cdot|s)$ and $v^2(\cdot|s)$, respectively. We obtain

$$V(v(p|s)) = \int_S U(q(s), p)[\,\beta v^1(p|s) + (1-\beta)v^2(p|s)]\, dp$$

$$= \beta \int_S U(q(s), p) v^1(p|s) dp + (1-\beta) \int_S U(q(s), p) v^2(p|s) dp$$

$$\leq \beta \int_S U(q(s), p) v^1(p|s) dp + (1-\beta) \int_S U(q(s), p) v^2(p|s) dp$$

$$= \beta V(v^1(p|s)) + (1-\beta)V(v^2(p|s)).$$

The inequality holds because $q^1(s)$ and $q^2(s)$ maximize expected utility if the posterior belief is given by $v^1(\cdot|s)$ and $v^2(\cdot|s)$, respectively. We have shown that the value function is convex in the posterior belief. The claim in the proposition then follows from Lemma 1.

Proposition 3 captures the direct welfare effect resulting from market transparency: since, by assumption, agricultural price risk cannot be hedged, the allocation of risk remains unaffected when the agricultural and food market becomes more transparent. All farmers benefit from higher market transparency because their exposure to price risk is lower at the time when they make their production decisions. It can be shown that transparency does increase ex ante expected output when the cost function is convex.

However, things are different when price risk can be hedged. While the direct welfare effect continues to be operative, an indirect welfare effect emerges: higher transparency destroys some risk-sharing opportunities and thereby imposes welfare costs on risk-averse farmers. If farmers are strongly risk-averse, the adverse indirect welfare effect dominates the favorable direct welfare effect and overall welfare decreases with higher market transparency. The normative implications of our economic issue for the agricultural and food sector thus depend on the risk-sharing opportunities available in the agri-food system.

Conclusions

In this paper, we have studied the impact of changes in market transparency on agricultural production levels. In particular, we have asked how the ex ante expected volume of agricultural production responds to additional information when information also affects the futures market for agricultural commodities. The analysis has produced two main results. Firstly, higher market transparency increases farmers' expected profits as long as the cost function is convex. However, higher market transparency does not increase agricultural production levels per se. If farmers' marginal cost functions are concave (convex), higher market transparency leads to higher (lower) ex ante expected levels of agricultural production.

Secondly, economic interpretation reveals that the value of additional information depends on two opposing effects. When farmer receive more reliable information they are able to improve their decisions, thereby increasing ex ante expected utility. However, additional information may interfere with the operation of risk-sharing markets thereby destroying some risk-sharing opportunities. Since farmers are risk-averse, ex ante expected utility declines. Due to these opposing effects, the overall impact of higher market transparency on farmers' ex ante expected utility remains ambiguous. Accordingly, when we vary the prevailing risk-sharing regime and assume that farmers have no access to futures markets or other risk-sharing opportunities, higher market transparency indeed does increase expected utility and leads to higher agricultural production levels.

Competing interests

The authors declare that they have no competing interests.

Authors' contributions

All authors read and approved the final manuscript.

Acknowledgments

We would like to thank our anonymous referees for helpful comments and suggestions.

Author details
[1] School of International Studies, Technische Universität Dresden, Dresden, Germany. [2] Department of Business and Economics, School of International Studies, Technische Universität Dresden, Helmholtzstr. 10, 01069 Dresden, Germany. [3] Department of Economics, University of Bielefeld, Bielefeld, Germany.

References
Allen DW, Lueck D (2003) The nature of the farm: contracts, risk, and organization in agriculture. MIT Press, Cambridge, London

Blackwell D (1953) Equivalent comparison of experiments. Ann Math Stat 24: 265–272

Broll U, Welzel P, Wong KP (2013) Price risk and risk management in agriculture. Contemp Econ 7: 17–20

Eckwert B, Zilcha I (2001) The value of information in production economies. J Econ Theory 100: 172–186

Eckwert B, Zilcha I (2003) Incomplete risk sharing arrangements and the value of information. Econ Theory 21: 43–58

Frechette DL (2000) The demand for hedging and the value of hedging opportunities. Am J Agric Econ 82: 897–907

Hirshleifer J (1971) The private and social value of information and the reward to incentive activity. Am Econ Rev 61: 561–574

Hudson D (2007) Agricultural markets and prices. Blackwell, Malden and Oxford

Kawai M, Zilcha I (1986) International trade with forward futures markets under exchange rate and price uncertainty. J Int Econ 20: 83–98

Kihlstrom RE (1984) A Bayesian exposition of Blackwell's theorem on the comparison of experiments. In: Boyer M, Kihlstrom RE (eds) Bayesian Models of Economic Theory. Elsevier, North-Holland, Amsterdam, pp 13-31

Moschini G, Hennessy DA (2001) Uncertainty, risk aversion, and risk management for agricultural producers. In: Gardner B, Rausser G (eds) Handbook of Agricultural Economics (Vol. 1a). Elsevier, Amsterdam, pp 88-153

Newbery DMG, Stiglitz JE (1981) The theory of commodity price stabilization. Clarendon Press, Oxford

UNCTAD (2011) Price formation in financialized commodity markets: The role of information, New York and Geneva

Wong KP (2007) Operational and financial hedging for exporting firms. Int Rev Econ Finance 16: 459–470

3

An empirical analysis on technophobia/technophilia in consumer market segmentation

Adele Coppola[*] and Fabio Verneau

* Correspondence: coppola@unina.it
Department of Agriculture,
Agricultural Economics and Policy
group, University of Naples Federico
II, Via Università 100, 80055, Portici
(Na), Italy

Abstract

Many factors can affect the success of food product innovations. One such factor is the role played by consumer attitudes and psychological factors, especially the way consumers feel towards technology, their attitude towards risk, and the perceived relationship between nutrition and health. With a view to analysing these factors, this paper first identifies consumer groups using a technophobia/technophilia scale and then relates attitude to technology with purchasing behaviour regarding products which have a higher level of manipulation. A set of statements based on the psychometric scale proposed by Cox and Evans was administered to a sample of 355 individuals intercepted as they left supermarkets and hypermarkets. Principal component analysis and cluster analysis were applied to identify groups of homogeneous individuals with regard to the behaviour of the interviewees in relation to technology. Results show the presence of seven different groups, including a small group of convinced technophiles (13% of the sample). This group of early adopters can play an important role in promoting the use of innovative products, thereby contributing to a rapid increase in demand. Moreover, an important aspect was the result with respect to confidence attributed to the media in ensuring correct and unbiased information regarding new food technologies. Many of the respondents judged the media negatively in this respect. However, appropriate use of the media could be an important lever to counteract the attitude of caution or scepticism.

Keywords: Psychometric scale; Consumption choices; Food innovation

1. Background

Product innovation is a key feature in company development strategy. This especially holds true in the food industry where increasing competitiveness is occurring alongside a schizophrenic evolution of consumer demand where environmental, social and ethical concerns coexist with hedonism, interest in innovative food and alternative cuisines, and the spreading of new eating habits, such as vegetarianism and veganism, seems to occur alongside their traditional counterparts. As a fact, consumer demand appears to follow two divergent directions. On the one hand, there is an increasing demand for products with a high technological content which respond to consumer needs in terms of practicality of use, nutritional content and amount of specific molecules (functional products with low calorie content, enriched foods). On the other, consumers increasingly demand organic products, food that has natural requisites, which is produced with environmentally friendly techniques, as well as typical products

and locally grown products, for which knowledge of the production area and the underlying tradition are guarantees in themselves (Tenbült et al., 2005; Verhoog et al., 2003).

These different trends may be connected to the way in which consumers behave
when faced with technology. Such behaviour influences how successful a new product
can be: the failure rate of new products is known to be very high, with previous studies
reporting different failure rates according to the country concerned or to the definitions
adopted of "failure" and "success". The FAO background paper on "Food product innovation" (Winger and Wall 2006) reported that 50% of new food and drink products
on grocery shelves are removed within two years. During 2000 and 2001 around 50% to
67% of the new products were withdrawn within one year from the food retailing
shelves in Germany, and similar figures, in most recent years (2010–2012), are reported
with the reference to the US market (Mintel International Group Ltd 2013).

The factors of success as regards food product innovation have been extensively studied (Kleijnen et al. 2009; Betoret et al. 2011). Besides the role of food retailers and their
"shelf position policy" (Winger and Wall 2006), and some characteristics of the market
(existing market vs new market, competitive environment, potential size of the market),
Balachandra and Friar (1997) pointed out the higher/lower ability of the new product to
meet consumers' needs, the understanding of consumer wants and preferences, as well
as the effectiveness of market segmentation and analysis. While market segmentation
practices usually take into account socio-demographic features, the success of innovative foods can also depend on consumer attitudes and psychological factors, especially
the way consumers feel towards technology, their attitude towards risk, and the perceived relationship between nutrition and health (Frewer et al. 2011; Rollin et al. 2011).

The latter aspects are the specific objects of our analysis. This paper aims, first of all,
to ascertain whether it is possible to identify and characterise consumer groups using a
technophobia/technophilia scale and, secondly, to correlate behaviour in relation to
technology with consumption choices of products which have a different level of natu-
ralness/manipulation. The degree of technophobia/technophilia was determined by referring to a psychometric scale (Food Technology Neophobia Scale, FTNS) set up by
Cox and Evans 2008, based on a series of statements which aim to detect the individual's attitude towards risk, technology and science. The use of this scale is substantially
based on the hypothesis that there is a positive correlation between a phobic attitude
towards food with a high technological content, on the one hand, and a more general
aversion to novelties and little confidence in science, on the other (Cox and Evans
2008; Evans et al. 2010a; Evans et al. 2010b). This paper used the set of statements proposed by Cox and Evans applying a principal component analysis to the statement
scores obtained from a field survey and using a cluster analysis to identify groups of
homogeneous individuals with regard to the individual's attitude towards risk, technology and science. The results of the analysis were then correlated with the consumption
choices of some products, with a view to ascertaining the link between psychological
aspects and consumption behaviour.

Methods
2.1 Neophobia and psychometric scales
In recent years the food sector has shown a highly innovative dynamic, and product
innovation has become a strategic element for the competitiveness of food companies.

The launch of new products is not always associated with a performance in the markets that rewards the costs on R & D and in some cases highly innovative food in terms of the content of nutritional elements or of techniques used in processing is accepted with caution or is completely rejected by consumers. According to anthropologists, eating means "incorporating", carrying within one's own body (Fischler 1990), and this may entail a certain caution or also aversion regarding the introduction of new, unknown elements within a diet.

Aversion to new foods has been the subject of several studies aiming to identify which psychological and social variables affect more or less phobic behaviour related to food consumption and/or set up techniques and analytical instruments that might help identify segments of the population with a different degree of neophobia/neophilia in relation to foods and hence with a different propensity to purchase and consume innovative goods.

In the context of studies on consumer behaviour, various techniques have been proposed to summarise the emotional factors connected to consumption attitudes and the convictions of individuals (Siegrist et al. 2008). In particular, in the field of cognitive psychology, consumer science and marketing, so-called psychometric scales have been used due to their enhanced capacity to identify segments of the population which are more/less neophobic, with a view to identifying the "first adopters" of innovative products. Table 1 reports the main psychometric scales linked to the neophilia/neophobia dichotomy referring to the adoption of new technologies, confidence in science and acceptability of innovative foods. All the psychometric scales reported in the table use a set of statements on which interviewees have to express their level of agreement measured on a Likert scale[a].

One of the first attempts to adopt a psychometric scale to food sector in order to identify and measure the most important factors in risk perception resulting from food hazard and food technology is the Perceived Food Risk Index (PFRI) which was drawn up using as starting point the pioneering studies of Slovic (Slovic et al. 1986). Trust in Science Scale (TISS) is a six item scale which focuses on public attitudes toward controversial scientific research and technologies (Bak 2001). The Food Neophobia Scale (FNS) used by Pliner and Hobden (Pliner and Hobden 1992) is a psychometric scale very close to the FTNS which consists of 10 statements with a seven-point scale, from "strongly disagree" to "strongly agree", and which allows us to determine the level of food neophobia in relation to the greater/lesser willingness to try new foods. The FNS is a specific application to the food products of a previous scale (General Neophobia Scale – GNS) developed by the same authors. Also the GNS, which does not refer to specific product categories, consists of 10 items (Pliner and Hobden 1992).

Table 1 Main psychometric scales

Acronym	Full name	Scope of use	Authors
FTNS	Food technology neophobia scale	Measuring the degree of food neophobia in humans linked to food technologies	Cox, Cox and Evans (2008)
FNS	Food neophobia scale	Measuring the degree of food neophobia in humans linked to food	Pliner and Hobden (1992)
GNS	General neophobia scale	Measuring the trait of neophobia in humans	Pliner and Hobden (1992)
TISS	Trust in science scale	Measuring trust in science and technology	Bak (2001)
PFRI	Perceived food risk index	Measuring consumer perception of food risk	Fife-Schaw and Rowe (1996)

Cox ed Evans measured correlations between FTNS, FNS, GNS, TISS and two different 'Willing to Try' scales, highlighting the strength of the FTNS scale in respect to predictive ability (Cox and Evans 2008). The FTNS is a useful tool to assess the impact of uncertainty and risk perception in relation to new technologies upon the acceptability of particular food. The FTNS has mainly been applied with reference to specific food technologies or certain types of foods linked to the use of sophisticated, new technological processes (Cox and Evans 2008; Evans et al. 2010a). However, further empirical testing of Cox's scale may be useful, first of all in relation to a set of foods rather than food technologies and, secondly, for a basket of common products which may be ordered on the basis of more/less technological content. This work concentrates specifically on these aspects.

2.2 Objectives and methods

This paper aims at analysing the relationship between the way consumers feels towards technology and consumption choices of products with a different level of naturalness/ manipulation.

We therefore performed an empirical analysis by means of a field survey based on administering a questionnaire. The survey was carried out in the region of Campania (southern Italy) and the interviews involved a sample of 355 individuals interviewed after shopping at supermarkets and hypermarkets. The composition of the sample was fairly balanced in terms of gender (48.7% men, 51.3% women) and over half the interviewees lived in the province of Naples. Other information on the socio-demographic characteristics of the interviewees may be noted in Table 2, showing certain specific features of the sample which help to interpret the survey results. First, the sample has a high share of self-employed workers (42.5%) and graduates (42.8%). Secondly, almost 60% of the interviewees are under 45 years old and in over two-thirds of cases the number of household members is equal to three or more. As regards monthly income, around one-third of the sample fell into each of the €1,000-2,000 and €2,000-3,000 classes.

The survey detected the opinions of consumers regarding categories of food products with a different technological content and related such opinions to their attitude to technology. In the first part of the questionnaire the interviewees were asked to express their level of agreement with a series of general statements on the role, importance and any effects of technology in the food sector. These statements were based on the set of 13 psychometric questions validated by Cox and Evans (2008). In the second part of the questionnaire, the survey focused on six product categories (functional products, organic, typical, short-chain, ready frozen food and diet products) to investigate specifically the confidence that the interviewee has in each type of food, the characteristics that can best describe it, and propensity and frequency of purchase.

The analysis on collected data was carried out in three steps.

First, an exploratory analysis was performed to understand how consumers feel about the proposed food categories, that is whether they are confident in them and how much they associate each product to a set of attributes (naturalness, safety, nutrition content, sense of gratification, taste, respect of the environment). To ascertain whether or not confidence in a product is associated to the opinion on the presence of a certain attribute a chi-square test was then carried out. A chi-square test was also used to verify the relationship between trust in a product category and the purchasing behaviour.

Table 2 Socio-demographic characteristics of the sample interviewed

Province	No. of interviewees	Percentage	Household income	No. of interviewees	Percentage
Caserta	20	5.63	Up to 1000 euro	46	13.0
Benevento	26	7.32	Between 1000 and 2000 euro	122	34.4
Avellino	39	10.99	Between 2000 and 3000 euro	131	36.9
Salerno	76	21.41	Over 3000 euro	54	15.2
Napoli	194	54.65	No reply	2	0.6
Total	355	100.0	Total	355	100.0
Profession	**No. of interviewees**	**Percentage**	**Household members**	**No. of interviewees**	**Percentage**
Unemployed	8	2.30	1	34	9.6
Manager	15	4.20	2	73	20.6
Other	21	5.90	3	87	24.5
Employee	78	22.00	4	102	28.7
Not in workforce	81	22.80	5 and over	59	16.6
Self-employed	151	42.50			
Total	355	100.0	Total	355	100.0
Age class	**No. of interviewees**	**Percentage**	**Qualification**	**No. of interviewees**	**Percentage**
Up to 35 years	117	33.0	Primary	11	3.1
36 – 45 years	93	26.2	Lower secondary	56	15.8
46 – 55 years	94	26.5	Higher secondary	136	38.3
Over 55 years	51	14.4	Degree	152	42.8
Total	355	100.0	Total	355	100.0

A second step of the analysis aimed to classify the interviewees according to their attitude towards technology and risk, measured by means of the 13 psychometric questions proposed by Cox and Evans. In this analysis, using principal component analysis (PCA) the 13 psychometric questions were grouped so as to specify complex indicators that might identify the way individuals relate to certain issues. On the basis of these complex indicators, the interviewees were classified into groups which are characterised by a different attitude to technology.

In the third and last step, groups of consumers, identified in step two, were related to the purchasing behaviour to verify by means of a chi square test whether the different attitude towards technology and risk actually influences the purchasing frequency of food with a higher/lower manipulation level.

3. Results

Analysis of the responses shows clearly differentiated opinions between organic, short-chain and typical products, on the one hand, and the food categories undergoing greater handling/processing, on the other. Whereas the first three categories of products were trusted by over 80% of the interviewees, over 50% responded that they had no confidence in foods with greater technological content, with a percentage that reached 61% in the case of ready frozen foods (Figure 1).

The dichotomy observed in terms of confidence also emerges in the analysis of attributes associated to the various product categories when the issue is naturalness

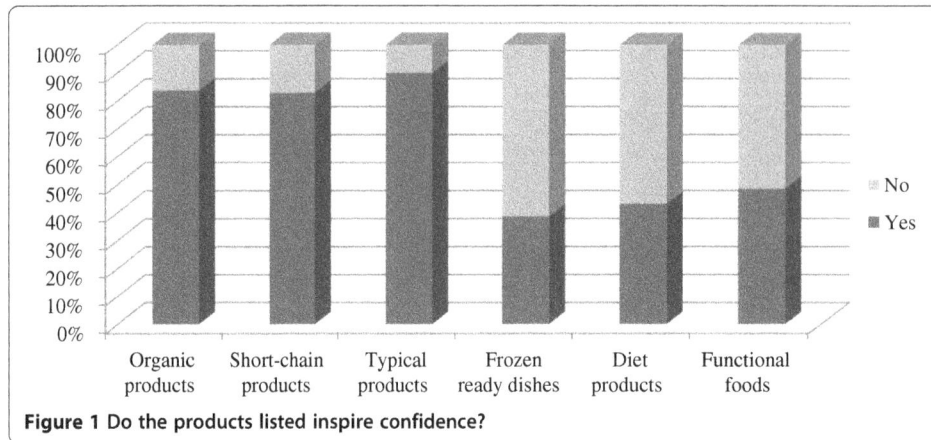

Figure 1 Do the products listed inspire confidence?

(Figure 2). Naturalness is the element which, according to the interviewees, clearly distinguishes organic, short-chain and typical products from the three other categories. Indeed, it is on the *naturalness* attribute that, as will be seen below, confidence in the former is based. The opinion on naturalness, which in this dichotomy appears to take on an ideological connotation, is reflected, at least for organic and short-chain products, in the responses concerning environmental respect. What is less clear, however, is the opinion on the other attributes in question, and it should be stressed that safety is associated to functional foods by a percentage of interviewees which does not differ greatly from that measured for organic products. Moreover, Figure 2 shows that consumers characterize each of the product categories with only few distinctive attributes: as an example, while organic products are mostly natural and safe, frozen ready dishes are mainly perceived as gratifying and tasty, and functional food are safe and nutritious.

To determine on which attributes the consumer bases the greatest/lowest trust in a product category, for each attribute we crossed the responses obtained as regards confidence with those concerning the most significant characteristics. The relations between the variables analysed were subjected to the chi-square test in order to ascertain whether or not confidence in a product is associated to the opinion on the presence of

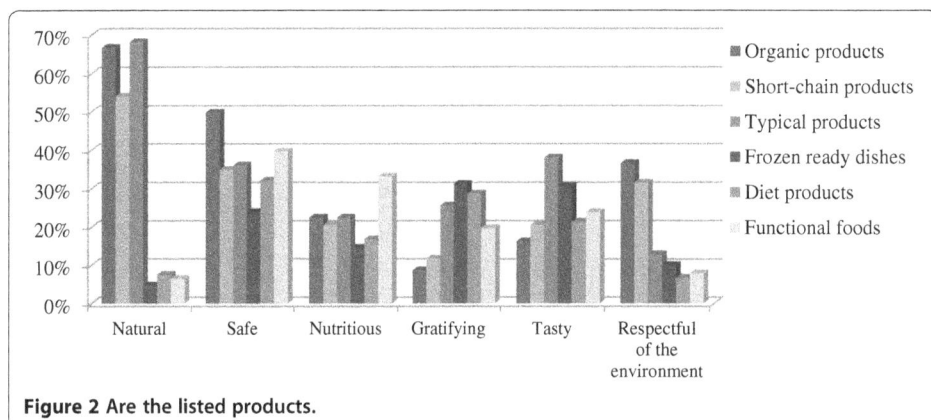

Figure 2 Are the listed products.

a certain attribute. The main results of the analysis are shown in Table 3 where values are reported only in case of significant relationships. What emerges is that:

- for organic products there is a significant relation between consumer confidence and their naturalness and safety. On analysing the specific data of the contingency tables, naturalness is shown to be a characterising element of organic products for 75% of those who trust such foods, against 26% who responded that they did not trust them;
- for functional foods the chi-square test shows a link between the presence or otherwise of confidence on the part of the interviewees and all the attributes investigated, except for the aspects that concern gratification and respect for the environment. The type of link may be made clearer by the percentages in question: 58% and 45% of those who trust functional foods believe that these products are, respectively, safe and nutritious;
- 73% of the interviewees who have confidence in typical products consider naturalness their distinctive element; only 22% of those who state they have no confidence in such products underline their naturalness as an attribute characterising this class of goods;
- a link between confidence and respect for the environment is statistically significant for short-chain and typical products, but this attribute seems to play a small role: only 34.4% of interviewees who trust in short-chain products state that they help the environment and the percentage is even smaller (11.9%) in the case of typical products;
- confidence and judgement on product safety are statistically related for all classes of food, but this attribute seems to be very important in case of organic, diet and functional products. Anyway, it should be underlined that for the last two products

Table 3 Confidence vs. attributes for various product categories by product category: contingence table*

		Are they natural?		Are they safe?		Are they nutritious?		Are they gratifying?		Are they flavoursome?		Do they help the environment?	
		No	Yes	No	Yes	No	Yes	No	Yes	No	Yes	No	Yes
Do you trust organic products?	No	74.1%	25.9%	75.9%	24.1%								
	Yes	25.3%	74.7%	45.1%	54.9%								
Do you trust short-chain products?	No	55.7%	44.3%	77.0%	23.0%							82.0%	18.0%
	Yes	43.9%	56.1%	62.6%	37.4%							65.6%	34.4%
Do you trust typical products?	No	77.8%	22.2%	77.8%	22.2%							77.8%	22.2%
	Yes	26.6%	73.4%	62.4%	37.6%							88.1%	11.9%
Do you trust frozen ready dishes?	No			79.8%	20.2%			65.1%	34.9%				
	Yes			70.1%	29.9%			74.5%	25.5%				
Do you trust diet products?	No	95.0%	5.0%	77.7%	22.3%	87.6%	12.4%						
	Yes	88.9%	11.1%	54.9%	45.1%	77.1%	22.9%						
Do you trust functional foods?	No	97.3%	2.7%	77.6%	22.4%	77.6%	22.4%			66.1%	33.9%		
	Yes	89.5%	10.5%	41.9%	58.1%	55.2%	44.8%			86.6%	13.4%		

*Only statistically significant relationships are included in the table.

the share of confident respondents is very low and safety assumes, then, a very slight role (see Figure 2).

Purchase frequency is statistically related to trust in the product (Table 4). This holds especially for functional foods as 42% of those who do not buy this type of product state that the lack of trust is the main motivation.

The attitude towards risk and technology can be inferred from the descriptive statistics in the Cox and Evans' scale (Table 5). Mean answers to the psychometric questions are very close to one another and always attain a score higher than 4, except for the perception of the effects on health in the long term and confidence in the media (3.69 and 3 respectively). The feeling of uncertainty seems to prevail in influencing the approach towards food technology. About 70% of the interviewees agree, to a greater or lesser extent, that new food technologies are something they are unsure of. This can be related to the fact that the benefits of new food technologies are often over-rated for 68.7% of the sample, that it may be risky to shift too hastily towards new food technology (58% of the sample), while only 36.6% of interviewees agree that new food technologies will not have negative effects on health in the long term.

Answers to psychometric statements are summarised by principal component analysis. The factor loading matrix (Table 6) reports the correlation coefficients between the 13 statements surveyed and the four components extracted in the PCA and allows the significance of the components to be ascertained. The first three express three different dimensions on which to construct perception of technology on the part of the consumer: riskiness, utility and the human possibility of controlling processes.

Table 4 Confidence vs purchase frequency by product category: contingence table and results of the chi-squared test

		Purchase frequency		chi-square	p-value
		Never or rarely	Often		
Do you trust organic products?	No	67,2%	32,8%	34,642	0,000
	Yes	27,3%	72,7%		
	Total	33,8%	66,2%		
Do you trust short-chain products?	No	55,7%	44,3%	8,816	0,003
	Yes	35,4%	64,6%		
	Total	38,9%	61,1%		
Do you trust typical products?	No	55,6%	44,4%	19,312	0,000
	Yes	21,9%	78,1%		
	Total	25,4%	74,6%		
Do you trust frozen ready dishes?	No	69,3%	30,7%	41,684	0,000
	Yes	34,3%	65,7%		
	Total	55,8%	44,2%		
Do you trust diet products?	No	69,3%	30,7%	30,716	0,000
	Yes	39,9%	60,1%		
	Total	56,6%	43,4%		
Do you trust functional foods?	No	71,6%	28,4%	54,172	0,000
	Yes	32,6%	67,4%		
	Total	52,7%	47,3%		

Table 5 Cox and Evans' scale: percentage of respondents and mean value

	1	2	3	4	5	6	7	Mean value
	Totally disagree (%)			Neither agree nor disagree (%)			Totally agree (%)	
It may be risky to shift too hastily towards new food technology	5.4	10.4	10.4	15.5	20.6	23.1	14.6	4.63
New food technologies in the long term may have negative effects on the environment	4.5	8.5	8.5	16.9	23.1	24.8	13.8	4.75
Society should not depend so greatly on new technologies to solve food issues	3.7	14.4	7.9	15.5	19.2	19.2	20.3	4.71
There are many tasty, nutritious foods around, so we do not need to use new food technologies to produce others	5.6	9.0	9.3	16.1	25.4	23.1	11.5	4.62
New food products are no more healthy than traditional foods	5.1	10.1	4.8	16.3	27.0	24.5	12.1	4.72
The benefits of new food technologies are often over-rated	2.0	7.3	7.9	14.1	28.5	25.9	14.4	4.95
New food technologies are something I'm unsure about	4.8	4.2	9.3	11.0	33.5	23.7	13.5	4.89
New food technologies diminish the natural quality of food	5.4	14.4	4.5	13.5	20.0	26.8	15.5	4.62
I have no reason to try highly technological foods because those that I eat are already good enough	5.6	9.0	9.3	16.1	25.4	23.1	11.5	4.89
New food technologies ensure that we all have greater control over food choices	5.4	13.2	16.9	17.7	24.5	16.6	5.6	4.15
The products obtained with new food technologies may help people to follow a balanced diet	5.1	14.6	12.1	18.0	24.8	18.9	6.5	4.25
New food technologies will not have negative effects on health in the long term	9.9	23.9	9.6	20.0	22.8	8.5	5.4	3.69
The media usually give correct impartial news on new food technologies	21.1	35.5	13.2	4.5	10.4	8.2	7.0	3.00

The first component explains 25.7% of the total variance and is correlated positively with the statements "It may be risky to shift too hastily towards new food technology" and "New food technologies in the long term may have negative effects on the environment". The component identifies the link between technology and risk perception (Siegrist 2008): introduction of new technologies in the food sector is interpreted as a factor aggravating existing risks, and the component which gradually shifts from negative to positive values captures the degree of risk aversion. This significance is also underlined by another question which has a high correlation index (even if it comes more into play in the second component) and which associates an idea of uncertainty to the new technologies. An important contribution to the first component is made by two other questions. The first expresses the opinion that society should not depend on new technologies to solve food issues. This statement may also be connected to risk perception: food safety is a strategic objective for society and should not depend on

Table 6 Factor loading matrix

	Components			
	1	2	3	4
	Risk perception	Perception of futility of technologies	Perception of benefits	Trust in the role of information
It may be risky to shift too hastily towards new food technology	0.777	0.122	0.003	−0.141
New food technologies in the long term may have negative effects on the environment	0.707	−0.045	−0.144	0.026
Society should not depend so greatly on new technologies to solve food issues	0.669	0.145	0.123	−0.027
There are many tasty, nutritious foods around, so we do not need to use new food technologies to produce others	0.479	0.353	−0.341	0.114
New food products are no more healthy than traditional foods	−0.136	0.736	0.049	−0.107
The benefits of new food technologies are often over-rated	0.141	0.673	−0.120	−0.042
New food technologies are something I'm unsure about	0.447	0.563	0.039	0.124
New food technologies diminish the natural quality of food	0.439	0.514	−0.258	−0.003
I have no reason to try highly technological foods because those that I eat are already good enough	0.315	0.510	−0.067	0.395
New food technologies ensure that we all have greater control over food choices	0.026	−0.091	0.792	0.080
The products obtained with new food technologies may help people to follow a balanced diet	−0.054	−0.152	0.689	0.250
New food technologies will not have negative effects on health in the long term	−0.071	0.193	0.621	−0.432
The media usually give correct, impartial news on new food technologies	−0.136	0.002	0.167	0.804

Kaiser–Meyer–Olkin Measure of Sampling Adequacy = 0.746.
Bartlett's Test of Sphericity = 853. df = 70; $p < 0.0001$.
Explained variance 56%.

technologies that are considered unreliable. The other statement "There are many tasty, nutritious foods around, so we do not need to use new food technologies to produce others" shows an aversion to technology in general and adds an ideological connotation to the first component.

The high levels of uncertainty and scarce perception of tangible benefits may generate in consumers the idea that new food technologies are superfluous. This is condensed in the second factor (12.2% of total variance) which captures the opinion of the interviewees as regards the usefulness of adopting new food technologies, and in general captures the level of uncertainty connected to them (*perception of futility of technologies*). When it takes on positive values, the second component identifies consumers who do not recognise a real benefit in the introduction of new technologies in food production. This approach is further reinforced by the feeling of uncertainty which goes with it ("New food technologies are something I'm unsure about") (Hansen et al. 2003; Frewer and Salter 2003). By contrast, negative values of the component are linked to the recognition by the interviewees that new technologies actually bring about a benefit in terms of health, food taste and quality.

The positive contribution that new technologies may make is better summarised in the third component (*perception of benefits*, explaining 9.5% of variance), correlated to the perception of benefits that new technologies have in terms of control over food choices, their capacity to ensure a balanced diet and the fact that they lead to positive effects on health (Weber et al. 2002).

The fourth and last component (8.5% of total variance) deviates from the others and puts the stress primarily on the role that the media may play in giving balanced and not biased information (*trust in the role of information*). This role appears particularly important as the component is also linked with the opinion on the effects which the new technologies may have on health. As one shifts from negative to positive values of the component, the trust in the media thus rises, as does the idea that in the long term new food technologies will have negative effects on health. This link, on one side, is coherent with the role media have played in many Italian food scandals, on the other side suggests how media are able to build a confidence feeling more on "bad news" than on information by itself.

In the second phase of the analysis the components extracted, starting from the psychometric questions, were used to classify the interviewees on the basis of their overall attitude to food technologies. Based on a hierarchical classification technique and choosing the number of groups where there is the maximum gap[b], we identified seven groups whose characteristics may be traced by looking at the centroid, in other words the average value which the components assume for each of them (Table 7), and, for a greater detail, at the mean scores attributed to the psychometric questions (Table 8).

Tables 9 and 10 report schematically the statements and characterising elements of the various groups. The first distinction among the interviewees in the groups may be made in relation to the degree of risk perception (component 1). In respect of this element, we may separate groups 3 and 6, which are both distinguished by negative values for this component, from the other groups for which the degree of risk perception is gradually increasing. Thus, except for groups 3 and 6, the sense of uncertainty and concern over the effects that new technologies could have in the future evidently prevails, to a greater or lesser extent, in all the interviewees.

Only for Group 3 (13% of the interviewed sample) can one speak of an actual technophile attitude[c] as a low perception of risk goes together with the recognition of the objective benefits in terms of food quality and contribution to diet. In Group 6 (17% of the sample), although there is no underlying concern or ideological preclusion in respect of food technologies, there emerges substantial scepticism on the utility of product

Table 7 Cluster analysis results

	Size of group	Risk perception	Perception of futility of technologies	Perception of benefits	Trust in the role of information
Group 1	14	1.190	−1.741	1.614	0.923
Group 2	59	0.551	0.641	−0.921	0.481
Group 3	48	−1.096	−0.814	0.303	−0.757
Group 4	95	0.185	0.133	0.115	−0.762
Group 5	32	0.675	−1.291	−1.313	0.043
Group 6	60	−0.951	0.252	−0.029	1.080
Group 7	47	0.454	0.833	1.064	0.025

Group size (no.) and centroids per group.

Table 8 Mean value of answers to psychometric questions by group

	Group						
	1	2	3	4	5	6	7
It may be risky to shift too hastily towards new food technology	6.0	5.5	2.9	5.1	5.3	3.0	5.6
New food technologies in the long term may have negative effects on the environment	5.1	5.4	3.4	5.2	5.8	3.6	5.2
Society should not depend so greatly on new technologies to solve food issues	6.3	5.5	3.6	5.0	4.4	3.8	5.2
There are many tasty, nutritious foods around, so we do not need to use new food technologies to produce others	4.2	6.0	2.9	4.5	5.4	4.0	5.2
New food products are no more healthy than traditional foods	2.1	5.5	4.0	5.1	2.3	5.0	5.8
The benefits of new food technologies are often over-rated	2.8	5.6	4.1	5.0	4.5	5.0	5.8
New food technologies are something I'm unsure about	5.0	5.9	3.3	5.0	4.0	4.8	5.7
New food technologies diminish the natural quality of food	2.5	6.2	2.7	4.8	5.1	4.4	5.5
I have no reason to try highly technological foods because those that I eat are already good enough	4.6	6.2	3.1	4.8	4.3	5.1	5.6
New food technologies ensure that we all have greater control over food choices	6.4	2.9	4.5	4.2	2.7	4.2	5.6
The products obtained with new food technologies may help people to follow a balanced diet	6.5	3.4	4.5	3.8	3.4	4.6	5.4
New food technologies will not have negative effects on health in the long term	4.3	2.2	4.4	4.6	1.9	3.0	4.9
The media usually give correct impartial news on new food technologies	4.7	3.1	2.2	1.9	2.4	5.0	3.2

innovations. By contrast, a convinced technophobe attitude is found in Group 2 (16% of interviewees) in which a high perception of risk and concern with the long-term effects on health and the environment accompany the idea of the futility of technology and the absence of benefits for diet and food choices. In the other groups these elements are combined with different intensities and make cautious attitudes emerge, yet which are more or less open to new technologies due to their ascribed benefits or the quality of the food or health aspects linked to them.

A particular aspect that should be stressed concerns the degree of trust attributed to the media in ensuring the new food technologies are covered correctly and impartially (component 4). Many of the interviewees expressed themselves negatively as regards the media, except for Groups 1 and 6 who believe that in the field of food technologies the media play a major role in information terms. Importantly, information could be an important lever to act on the attitude of caution or scepticism as regards new products which characterises these two groups, that accounts overall for one fifth of the sample.

Confirmation of the results of the group analysis was sought by linking the extreme groups[d] (the convinced technophobes and the convinced technophiles) to purchase of products with a higher level of manipulation, that is, functional, light and ready to eat and frozen products. Indeed, extreme groups are those where the mean values of all the principal components (the centroids) have opposite signs, and thus attitude to food technology should be better defined. For the extreme groups Table 11 reports the percentage of people purchasing, or otherwise, the three food categories. In all cases the convinced technophobes show a higher, and statistically different, percentage of non-buying interviewees with respect to the convinced technophile group. This particularly holds for diet products where 94% of the group states that they buy them, while the percentage reaches only 58% within the convinced technophobes.

Table 9 Characterising statements by group

Group	Characterising aspects	Definition
Group 1	They perceive in great measure the risk and uncertainty associated to new technologies. Yet, at the same time, they think that products obtained with new technologies may have beneficial effects on diet and allow control of food choices. They do not believe that in the long term product innovations can have negative effects on health. They trust the role that the media perform in supplying sound information. Almost 64% of the group have children under 12 years old.	Cautious
Group 2	There is a strong perception of risk and uncertainty associated to new technologies, but what prevails is the idea of their uselessness because they do not provide more tasty or healthy products, nor do they allow the diet to be controlled. They believe that in the long run new food technologies may have negative effects on health and the environment.	Convinced technophobes
Group 3	They do not consider food technology risky and do not associate uncertainty to it. They consider product innovations useful because they lead to tasty, better-quality food and allow greater control over diet. They do not think that, in the long term, technology can be detrimental to health. They have no confidence in the informational role of the media. Over 60% of Group 3 have a university degree.	Convinced technophiles
Group 4	They represent the mid-point of the sample and summarise commonly-shared opinions on new food technology. Worried about the effects on health and the environment, they believe that traditional foods are better and more healthy, and that the benefits of the new technology are overrated. They are distinguished by their total lack of confidence in the ability of the media to supply correct, impartial information on new food technologies. Only 28% of the group have a university degree.	Traditionalists
Group 5	They have a high perception of the risk associated with food technologies and are concerned at the negative effects they might have on the environment and health. They have no confidence in the ability of new foods to contribute to a more balanced diet, but admit that new products may be more tasty and healthy. They have no confidence in the ability of the media to provide reliable information on such technologies.	Tendency to be technophobic
Group 6	They are not ideologically opposed to new food technologies: they do not consider it risky to adopt new technologies too hastily and are not concerned by dependence upon them to solve food problems. However, they do not think that innovations in the food sector are useful. They believe their benefits are overestimated and think it is possible to find products that are already quite good. In general, there are uncertain regarding new technologies and fear negative long-term effects on health. Males make up 68% of the group.	Sceptics
Group 7	This group perceives the benefits that technological foods may confer in terms of controlling food choices and achieving a balanced diet. Nevertheless they consider new food technologies risky and essentially futile as they think that they diminish the natural quality of food and that there are already good, healthier products on the market. Females make up 63.8% of this group.	Tendency to be technophile

Table 10 Socio-demographic statistics by group

Group	Females (%)	Graduates (%)	Families with chidren (%)
Cautious	57.1	57.1	64.3
Convinced technophobes	47.5	50.8	30.5
Convinced technophiles	41.7	62.5	22.9
Traditionalists	63.2	28.4	23.2
Tendency to be tecnophobic	53.1	40.6	28.1
Sceptics	31.7	35.0	33.3
Tendency to be technophile	63.8	48.9	19.1

Table 11 Purchasing behaviour in convinced technophobe/technophile groups

Do you buy?...		Convinced technophobes	Convinced technophiles	χ^2	p-value
Frozen ready dishes	No	30.5%	8.3%	7.968	0.005
	Yes	69.5%	91.7%		
		100.0%	100.0%		
Diet products	No	42.4%	6.3%	17.875	0.000
	Yes	57.6%	93.7%		
		100.0%	100.0%		
Functional food	No	35.6%	16.7%	4.799	0.028
	Yes	64.4%	83.3%		
		100.0%	100.0%		

4. Discussion and conclusions

The relationship between attitudes, intentions and behaviour is extensively covered in the literature and has been one of the chief aspects tackled by consumer studies in the economic and social field for the past 25 years. The addition of attitudinal and psychographic factors to the socio-demographic variables traditionally used by economic theory has led to the progressive development of psychometric scales to capture conceptual dimensions which are able to interpret value systems, lifestyles and attitudes, and translate them into measurable variables.

The use of psychometrics in the social sciences, economics and marketing has at times focused on the dichotomy between technophobe and technophile attitudes in the food sector. The prevalence of one or the other may greatly affect the choices and performance of the food industry interested in R&D investment to introduce highly innovative products onto the market. Identification and quantification of the market segments most attracted by innovative products, combined with analysis of the socio-demographic, behavioural and psychographic characteristics in such groups, may steer company choices both in terms of investment and communication strategies.

Our study falls within the above research strand. The aim was to segment a sample of consumers in the southern Italian region of Campania on the basis of their degree of technophobia in the food sector. Our analysis was conducted using the psychometric scale proposed by Cox and Evans, the Food Technology Neophobia Scale (FTNS), applied to six categories of food products which had undergone various degrees of industrial handling and processing. The variables generated by the FTNS were subjected to principal component analysis (PCA) which extracted four factors subsequently used for sample segmentation via an analysis of non-hierarchical groups.

The results show the separation of seven groups distinguished by different attitudes to food technology, determined by their perception of risk, perception of uncertainty and by personal judgements concerning the utility, benefits and potential harm of using such technology. There emerged a group which we called *convinced technophiles* consisting of 48 individuals, about 13.5% of the sample. The group represents so-called early adopters who are the first ones taking part in the market for food products with a high innovation content[e]. Starting from this group, other consumers, or *followers*, may take up the demand, allowing the firm to penetrate the market and achieve their objectives in terms of sales and market share (Mahajan et al. 2000).

The availability of new techniques to refine market segmentation may lead to more effective marketing policies being chosen by the company, allowing both a clearer view of the different life cycle stages of the product itself and offering the possibility to affect it especially in regard to the crucial phases of introduction and product growth, during which food companies seek to recover the costs of R&D (product concept and design) and those carried out to start production. From this point of view the group of the early adopters seems to play a strategic role: knowing not only their socio-demographic characteristics, but also psychographic profiles represents a crucial factor for the firm both to estimate potential demand and to target choices in terms of distribution channel and communication.

Endnotes

[a]The Likert Scale is a psychometric ordered scale commonly used in surveys and questionnaires to measure opinions and attitudes. In a Likert scale the respondent is presented with a set of attitude statements on a scale ranging from strongly agree to strongly disagree and he/she chooses one option that best aligns with his/her view.

[b]Number of the groups is chosen looking at the distance index in the clustering process.

[c]Actually, the use of self-anchoring scales allows to get information on the absence/presence of technophobia. However, in different studies, the absence of neophobia is interpreted as the presence of neophilia. Look in this regard the work of Choe and Cho (2011) that using the scores of the FNS divides the participants in the survey in neophilics and neophobics. The same approach is also used in Tuorila et al. (2001).

[d]Cluster analysis classifies individuals on the basis of similarities of some characteristics seeking to minimize within-group variance and maximize between-group variance. A certain level of variance within groups is "physiological", especially due to factors that are less relevant in characterizing the group. That is why we chose to show the results for the extreme groups where the distribution of components 1, 2 and 4 is clearly defined around negative (group 3) or positive (group 2) values and the opposite is true for component 3. That allows to better relate purchase behavior to the attitude toward risk, uncertainty, confidence in media and to draw up conclusions that can be true for the group as a whole.

[e]Many authors argue that individuals characterized by lower aversion toward food technology can represent a segment of early adopters. Cox and Evans (2008) explicitly link the degree of neophobia with the possibility to identify segments of the early adopters and more recently other authors have used tecnophilia-tecnophobia as a predictor of early adoption (Popa and Popa 2012).

Competing interests
The authors declare that they have no competing interests.

Authors' contributions
FV carried the study plan, collected data, and wrote the background and the method paragraphs. AC analyzed the results, and wrote the results and discussion and conclusion paragraphs.

Acknowledgements
The authors are grateful for the valuable comments of the anonymous reviewers.

References

Bak H (2001) Education and public attitudes toward science: implications for the "Deficit Model" of education and support for science and technology. Soc Sci Q 82(4):779–795

Balachandra R, Friar JH (1997) Factors for success in R&D projects and new product innovation: a contextual framework. IEEE Trans Eng Manage 44(3):276–287

Betoret E, Betoret N, Vidal D, Fito P (2011) Functional foods development: trends and technologies. Trends Food Sci Technol 22(9):498–508

Choe JY, Cho MS (2011) Food neophobia and willingness to try non-traditional foods for Koreans. Food Q Prefer 22:671–677

Cox DN, Evans G (2008) Construction and validation of a psychometric scale to measure consumer's fears on novel food technologies: the food technology neophobia scale. Food Q Prefer 19:704–710

Evans G, de Challemaison B, Cox DN (2010a) Consumers' ratings of the natural and unnatural qualities of foods. Appetite 54(3):557–563

Evans G, Kermarrec C, Sable T, Cox DN (2010b) Reliability and predictive validity of the Food Technology Neophobia Scale. Appetite 54(2):390–393

Fife-Schaw C, Rowe G (1996) Public perceptions of everyday food hazards: a psychometric study. Risk Anal 16(4):487–500

Fischler C (1990) L'Homnivore: le goût, la cuisine et le corps. Odile Jacob, Paris

Frewer LJ, Bergmann K, Brennan M, Lion R, Meertens R, Rowe G, Siegrist M, Vereijken C (2011) Consumer response to novel agri-food technologies: implications for predicting consumer acceptance of emerging food technologies. Trends Food Sci Technol 22(8):442–456

Frewer LJ, Salter B (2003) The changing governance of biotechnology: the politics of public trust in the agri-food sector. Appl Biotechnol Food Sci Policy 1(4):199–211

Hansen J, Holm L, Frewer L, Lynn J, Robinson P, Sandoe P (2003) Beyond the knowledge deficit: recent research into lay and expert attitudes to food risks. Appetite 41:111–121

Kleijnen M, Leeb N, Wetzels M (2009) An exploration of consumer resistance to innovation and its antecedents. J Econ Psychol 30(3):344–357

Mahajan V, Muller E, Wind Y (ed) (2000) New-Product Diffusion Models. Kluwer Academic Press, Boston & Dordrecht

Mintel International Group Ltd (2013) The NASFT State of the Industry Report – The Market

Popa ME, Popa A (2012) Consumer behavior: determinants and trends in novel food choice. Novel Technol Food Sci 7:137–156

Pliner P, Hobden K (1992) Development of a scale to measure the trait of food neophobia in humans. Appetite 19(2):105–120

Rollin F, Kennedy J, Wills J (2011) Consumers and new food technologies. Trends Food Sci Technol 22(2–3):99–111

Siegrist M (2008) Factors influencing public acceptance of innovative food technologies and product. Trends Food Sci Technol 19(11):603–608

Siegrist M, Stampfli N, Kastenholz H, Keller C (2008) Perceived risks and perceived benefits of different nanotechnology foods and nanotechnology food packaging. Appetite 51(2):283–290

Slovic P, Fischhoff B, Lichtenstein S (1986) The psychometric study of risk perception. Risk Eval Manage 1:3–24

Tenbült P, de Vries NK, Dreezens E, Martijn C (2005) Perceived naturalness and acceptance of genetically modified food. Appetite 45(1):47–50

Tuorila H, Lähteenmäki L, Pohjalainen L, Lotti L (2001) Food neophobia among the Finns and related responses to familiar and unfamiliar foods. Food Q Prefer 12(1):29–37

Verhoog H, Matze M, Lammerts van Bueren E, Baars T (2003) The role of the concept of the natural (Naturalness) in organic farming. J Agric Environ Ethics 16(1):29–49

Weber EU, Blais A, Betz NE (2002) A domain-specific risk-attitude scale: measuring risk perceptions and risk benefit. J Behav Decis Mak 15(4):263–290

Winger R, Wall G (2006) Food product innovation: a background paper. Agricultural and food engineering working document 2. FAO, Rome

4

Factors influencing farmers' behavior in rice seed selling in the market: a case study in the Tarai region of Nepal[d]

Narayan P Khanal[*] and Keshav L Maharjan

* Correspondence:
narayankhanal36@gmail.com
Graduate School for International
Development and Cooperation
(IDEC), Hiroshima University, 1-5-1
Kagamiyama, Hiroshima 739-8529,
Japan

Abstract

The importance of rice in food security and livelihoods of Nepalese people is well recognized but the seed supply system of this crop in the rural areas is poorly developed. To increase farmers' access to a wide range of rice varietal choices in a cost effective way, some farmers, organized in groups or cooperatives, have started producing and marketing rice seed through development projects in recent years. But very limited information has been published about the performance of the farmers in rice seed marketing. In this study, we analyze the impact of households' socio-economic variables on market participation and volume of rice seed sold, using a Heckman selection model. Data for the study were collected from three *Tarai* districts (Siraha, Kailali and Chitwan) in Nepal with a sample size of 180, that is, 60 households from each district. Result shows that 65% of households sell 64% of rice seed produced. Households with agricultural training, share contribution to their organization and higher livestock numbers are more likely to participate in the market. Similarly, households with older household heads, higher operational land, and access to an irrigation facility sell a higher amount of rice seed in the market. Seed price has a positive influence both on market participation and seed volume sold.

Keywords: Farmers' behavior; Rice seed selling; Market participation; Heckman selection; Nepal

Background

Rice is the most important food crop in Asia as it contributes 60% of households' calorie consumption, and about 90% of the world's rice output is produced and consumed in the continent (FAOSTAT 2012). Rice provides 20% of agricultural gross domestic product, 40% of calorie and 25% of protein requirements in the Nepalese diet from cereals (MoAC 2010a). In spite of this, rice yield in Nepal is 2.9 t ha^{-1}, which is lower than that of neighboring countries (India-3.5 t ha^{-1}, Bangladesh-4.4 t ha^{-1}, China-6.6 t ha^{-1}), and the world average (4.2 t ha^{-1}) (FAOSTAT 2012). In addition, there is high (50%) disparity between potential and national average rice yield in this country. One of the reasons for this is poor access to quality seeds by farmers, which is evident from the poor seed replacement rate (SRR)[a]. The recommended SRR for self-pollinated crops including rice is 25% but its SRR was only 8.7% in Nepal in 2010. This means only 8.7% of the total rice area received quality seed in that year, and the rice seed required for the remaining area was supplied from informal sources such as rice grain reserved from previous season's harvest or local exchange of rice seed with neighbours (SQCC 2011). Also, the share of the government corporation also known as 'National Seed Company' in rice seed supply in 2010 was only 17%, and rest of the seed was supplied through development projects, farmers' groups, cooperatives, and agrovets–small traders supplying agricultural inputs (SQCC 2011; Pokhrel 2012). The involvement of private companies in rice seed supply is quite low (<5%), which might be due to low profit margin and high fluctuation of rice seed demand in the market (Almekinders et al. 1994; Lal et al. 2009; SQCC 2011). In Nepal, rice is grown from *Tarai* (from 70 m to 650 m above mean sea level-amsl) to hills (up to 3,050 m amsl-the highest rice growing altitude in the world), with the dominance of small farmers (average land holding 0.8 ha per household with 50% of households have <0.5 ha) (CBS, 2003). Also, variation exists across the production plots of the same household in terms of irrigation facility and soil characteristics (MoAC 2010a). To address opportunities and constraints associated with the above-mentioned variations together with climate and market factors, farmers tend to diversify their rice varietal portfolio using both modern (those developed from research organisations) varieties and local landraces (Gauchan et al. 2005; Khanal & Maharjan 2013). This demands a mechanism that supplies a wide range of rice varieties to farmers in a cost effective way.

From the 1990s, government and Non-Government Organizations (NGOs) started facilitating farmers' groups and cooperatives (also called as community-based seed producer organizations-CBSPOs) in rice seed production and marketing through action research, and development projects (Lal et al. 2009; Witcombe et al. 2010). Government statistics show that 146 registered CBSPOs with 3,500 members (households) were involved in rice seed production in 2009. The majority of these registered CBSPOs (80%) are from the *Tarai* region, the food basket of Nepal, contributing 70% of the total rice produced in the country (MoAC 2010b). The CBSPOs have also been promoted in other countries of Asia and Africa to contribute in the local seed supply of various food crops including rice (David 2004; Bishaw & van Gastel 2008; Srinivas et al. 2010). It is believed that CBSPOs could produce and market rice seed in a cost effective way because these activities are handled by farmers at local levels. Also, farmers are in a better position than other actors such as government agencies and private companies to select appropriate rice varieties suitable for local niches due to their experience in rice farming and informal networks (Almekinders et al. 1994).

In spite of the great potential of CBSPOs in supplying seeds of different rice varieties at local level, it is not clear how small farmers could increase their participation in marketing (Almekinders et al. 1994; Setimela et al. 2004; Pokhrel 2012). Very few studies have been published on this issue (David 2004; FAO 2010; Srinivas et al. 2010; Witcombe et al. 2010), and the focus of these studies is on the ability of CBSPOs to cover marketing (processing, storage and distribution) costs and provide additional benefits to their members (farmers) using estimated seed production data. These studies might have limited policy implications as the estimated production data would be less likely to represent the actual volume of seed sold by households in the market. This is because in developing countries the cereal seed industry is in an early development stage, and farmers tend to sell part of their agricultural produce in the market and the remaining output is consumed at home, and/or exchanged with neighbours as seed or grain (Amekinders & Louwaars 1999). Since rice seed production is carried out at household level and marketing through their organization, household characteristics could play a vital role in farmers' performance in rice seed marketing. This article intends to measure the impact of households' socio-economic factors on rice seed marketing.

Literature review

Microeconomic theory explains the farmers' behaviors in selling seed in the market. According to theory, producers' and consumers' behaviors in the market varieties with market signals. It means the volumes of commodity supplied in the market chain is guided by price, and there is no barrier to entry and exist of enterprises in the market chain. However, it is difficult to explain farmers' behaviors in selling agricultural produce in subsistence farming considering only price signal, whereby farmers do not sell a significant portion of seeds produced (Almekinders & Louwaars 1999; Omti et al. 2009). Previous studies indicate that farmers' participation in marketing is constrained by productivity (Janvry et al. 1991; Fafchamps 1992), poverty, information, and physical constraints including road and storage facilities. Heltberg and Tarp (2002) found that farm size, animal manure, age of household head (HHH), ownership of transportation means and strategies to adapt to climate risk to be positively associated with market participation. All these variables are related to productivity enhancement. Similarly, Benfica et al. (2006) found a significant positive impact of education of household head, access to credit and households' off-farm income on market participation. Access to storage facility is another challenge for farmers in rice seed marketing (Barrett 2006). Despite the availability of some literature about the impact of households' socio-economic characteristics, local level and crop specific studies are needed to make appropriate policy recommendations.

Methods

Study site and sampling technique

A field survey was carried out in three districts (Sirha, Chitwan and Kailali) of the region of Taria in Nepal, representing Eastern, Central and Far-western development regions, respectively. A multi-stage random sampling method was used to select the sample households. In each district, first, CBSPOs having at least two years experience in rice seed production and marketing were selected in consultation with District Agriculture Development Offices (DADOs). Then, in the second step, 15 members from each of the selected CBSPOs were randomly selected, making the total sample size of 180 for the survey. After the completion of the household survey, meetings were held with the selected CBSPOs to discuss issues raised during household survey. The survey was implemented in October and November 2011.

Empirical model

A Heckman selection model (Heckman 1979) was used for data analysis. This model is used where sample selection arises as a result of partial observation of the outcome variable. In the presence of sample selection, the observed outcome does not represent a random sample of its population. In this case, ordinary least square (OLS) regression could give biased results as it nullifies the censored observations. The Heckman selection model has been developed to address this problem. This model consists of two equations. The first equation is called the selection equation which is similar to a probit model, and it measures the impacts of socio-economic variables on the probability of households' selling rice seed in the market. The second equation, also calledoutcome equation, is similar to an OLS model and measures the impact of socio-economic variables on volume of seed sold in the market. These two equations are modeled by a two-step procedure or a one-step procedure. Though the two-step procedure has been frequently used in the literature, the one-stage procedure (simultaneously modeling of two equations using the maximum likelihood method) would be more efficient than the two-step procedure (Nawata & Nagase 1996; Nawata 2004). Nawata (2004) argued that the one-step procedure is more appropriate than the two-step procedure if sample size is small. Considering these issues, we followed the one-step procedure in the study. Another consideration in the Heckman selection model is the use of an identifier variable which is used in selection equation and not in the outcome equation. We used rice seed price for this purpose. The outcome and selection equations are presented in equations 1 and 2, respectively.

$$y_i = x_i \beta + \mu_i \tag{1}$$

$$Z_i^* = w_i \alpha + \epsilon_i \tag{2}$$

where y_i is volume of seed sold in the market, Z_i^* is a latent variable, x_i and w_i are the vectors of explanatory variables, β is the vector of coefficients, and μ_i and ϵ_i are the error terms.

The operational models of the outcome and selection equations are given in equations 3 and 4, respectively.

$$\begin{aligned} \ln \text{seed sold} = {} & \beta_0 + \beta_1 \text{age of HHH} + \beta_2 \text{education of HHH} + \beta_3 \text{family labor} + \\ & \beta_4 \text{cultivated land} + \beta_5 \ln \text{off-farm income} + \beta_6 \text{irrigation} + \beta_7 \text{livestock} + \\ & \beta_8 \text{training} + \beta_9 \text{share} + \beta_{10} \text{roof type} + \mu_i \end{aligned} \tag{3}$$

$$\begin{aligned} \text{Market participation} = {} & \alpha_0 + \alpha_1 \text{age of HHH} + \alpha_2 \text{education of HHH} + \alpha_3 \text{family labor} \\ & + \alpha_4 \text{cultivated land} + \alpha_5 \ln \text{of-farm income} + \alpha_6 \text{irrigation} + \alpha_7 \text{livestock} + \alpha_8 \text{training} \\ & + \alpha_9 \text{share} + \alpha_{10} \text{roof type} + \alpha_{11} \text{seed price} + \epsilon_i \end{aligned}$$

$$\tag{4}$$

where ln is log. Seed sold is the dependent variable used in the outcome equation which indicates the quantity of rice seed sold by farmers in the market.

It is possible that farmers' sell seed not only to CBSPOs but also to other actors such as local farmers, agrovets, and development projects. However, CBSPOs and DADOs in group discussions argued that farmers in the study area rarely sell seed directly to other actors. Rather they sell seed to the CBSPOs where they have taken membership, and the CBSPOs after processing (packaging, quality checking and leveling) sell seeds to the aforementioned actors. So, we consider CBSPOs are the markets for farmers. Similarly, market participation is the dependent variable in the selection equation which shows whether farmers sell seed to CBSPOs or not (i.e. a dummy variable which takes the value 0 or 1).

A total of 11 socio-economic variables were chosen as explanatory variables considering economic theory, findings from previous literature and experience of farmers as the combination of these strategies would help to draw the relevant variables for the study (Table 1). These variables include demographic (age and education of HHH, and family labor), economic (cultivated land, irrigation facility, off-farm income, livestock and roof type), and institutional (training, and having a share in an CBSPO). The justification for the selection of these variables is given below.

The impact of age and education of HHH was hypothesized to be positive because age represents experience and education indicates analytical capability, both of which might have a positive impact on households' market participation and volume of seed sold (Heltberg and Tarp 2002). Similarly, rice seed production is carried out in rural areas where the majority of the work is done by the family members. Also, rice farming is seasonal in nature when most of labourers are busy in their own households' activities. Even those wanting to hire labourers might not get them on time and could not carry out field activities properly, which might influence the quantity and quality of rice seed. So, it was hypothesized that family labour (Labour force unit–LFU[b]) would have a positive impact on both market participation and seed sale volume.

Amount of cultivated land, irrigation facility (proportion of the total amount cultivated land with irrigation facility) and organic manure have a positive linkage with crop yield (Azam et al. 2012), so these variables were assumed to have a positive impact on the marketing indicators. We used livestock (Livestock Standard Unit–LSU[c]) as a proxy variable to represent the amount of animal manure applied in rice fields. Similarly, those having higher off-farm income might be less affected by cash/food shortage, especially the period between rice crop harvest and its seed sale, and would be more motivated towards marketing. Moreover, the CBSPOs might be poor in physical structure (e.g., storage house, grading machine) in the early phase of the cereal seed industry development. This implies that seed growers might have to store seed at their personal

Table 1 Description of variables and expected sign

Variables	Definition	Mean ± SD	Expected sign
Seed sold	Amount of rice seed sold by farmers (kg)	1,356.7 ± 144.3	
Seed selling	1 = if they sell the seed, 0 for otherwise	65.8 ± 0.47	
Age HHH	Age of HHH in years	46.83 ± 11.43	+
Education HHH	Formal schooling years of HHH	7.96 ± 4.02	+
Family labor	Labor force unit (LFU)[2] at household	3.44 ± 1.44	+
Cultivated land	Total operational land for rice seed production (ha)	0.95 ± 0.36	+
Off-farm income	Annual cash income of household members from off-farm sources (NRs)	42,998 ± 38,234	+
Irrigation	% operational land area under irrigation facility	54.5 ± 26.8	+
Livestock	Livestock standard unit (LSU)[1]	3.86 ± 5.77	+
Training	1 = if household received seed management training, 0 = otherwise	0.783 ± 0.413	+
Share	1 = If farmers put share in CBSPOs, 0 = otherwise	0.644 ± 0.480	+
Roof type	1 = if households have concrete roof and o = otherwise	0.338 ± 0.645	+
Seed price	Price of rice seed (NRs per kg)	18.02 ± 2.81	+

1 US$ = NRs. 82.96 (Nepal Rastra Bank, 2011.11.30).
Source: Field survey, 2011.

houses for a few months after the rice harvest until CBSPOs make arrangements to store it in their warehouses. Those having concrete-roofed houses would be more likely to be motivated towards marketing than their counterparts as they could store the seed maintaining quality.

Training and household's share holdings (cash deposited at CBSPOs by farmers) are the two institutional variables considered in the study. It was assumed that those receiving training in any aspect of seed management (production, quality control and marketing) might be better off both in the market participation and seed selling volume as training tends to enhance households motivation towards marketing. Similarly, those who deposited cash at CBSPOs as share were assumed to have better performance in marketing. It is because profit generated from the marketing of seed could be distributed to households based on the proportion of share amount they deposit at CBSPOs. The detail of dependent and explanatory variables used in the study is presented in Table 1.

Before running the Heckman selection model, data were validated for multicollinearity and heteroskedasticity. The Variance Inflation Factor (VIF) method was used to detect multicollinearity because this method is preferred over the correlation coefficient method (Pindyck & Rubinfield 1981). We did not find a problem of multicollinearity in the explanatory variables used in the model as the values are less than 10. The test for homogeneity of variance was conducted using the Breusch-Pagan/Cook-Weisberg test for heteroskedasticity, and the hypothesis of constant variances of the residuals was not rejected (p > 0.25) in both equations. Moreover, endogeneity issue was checked across the price variables in both the equations through Hausman test and did not find problem.

Results

Summary of the selected variables

The study shows that 65.8% of farmers sold rice seed in the market on average 1,356.7 kg household^{-1} and this volume is 64% of the total rice seed produced by households (Table 1).

The average amount of cultivated land for rice seed production per household was 0.95 ha, and this area represents 85% of the total amount of cultivated land 1.16 ha. Similarly, households make their livelihoods from various on-farm (cereal crop, vegetable, livestock, etc.), and off-farm (labor work, salaried job, small business and remittance) sources; however, the share of the latter sources to the annual households' average cash income is 69%. The average age of HHH was 46.83 years but it varied from 17.0 to 75.0 years. The average LSU was 3.86, and major animals raised by farmers include cows, buffaloes, goat and poultry. The majority of HHH in the study area received agricultural training (78.3%) from government organizations and NGOs. About one third of the households (33.8%) had concrete roofed houses. About two-third of the household (64.4%) adopted the practice of depositing shares in their organizations. The average price of seed was NRs 18.02 kg^{-1} but it varied from NRs 17 kg^{-1} to NRs 24 kg^{-1}.

Output from Heckman selection model

Table 2 presents the result from the Heckman selection model and shows that the variables chosen for the study fit this model well which is shown by the significant log

Table 2 The impact of explanatory variables on outcome and selection equations

Variables	Outcome equation		Selection equation	
	Coefficient	Marginal impact	Coefficient	Marginal impact
Age HHH	0.020 (0.034)**	0.019 (0.036)**	−0.002 (0.772)	−0.001 (0.773)
Education HHH	0.027 (0.401)	0.031 (0.342)	0.021 (0.496)	0.006 (0.493)
Family labour	0.0213 (0.254)	0.031 (0.402)	0.027 (0.498)	0.009 (0.503)
Cultivated land	0.07 (0.008)*	0.06 (0.048)**	0.005 (0.220)	0.0017 (0.229)
Off-farm income	0.4 (0.301)	0.2 (0.231)	0.1 (0.746)	0.1 (0.856)
Irrigation	0.0765 (0.072)*	0.042 (0.072)*	0.112 (0.795)	0.036 (0.794)
Livestock	0.004 (0.342)	0.002 (0.221)	0.09 (0.094)*	0.071 (0.048)**
Training	0.074 (0.76)	0.0212 (0.78)	0.182 (0.009)***	0.155 (0.014)**
Share	0.081 (0.815)	0.114 (0.309)	0.220 (0.037)**	0.190 (0.037)**
Roof type	0.271 (0.212)	0.259 (0.217)	0.033 (0.896)	0.010 (0.13)
Seed price	-	0.11 (0.084)*	0.08 (0.062)*	0.071 (0.045)**
Constant	6.433 (0.001)***		2.95 (0.008)***	

Wald Chi (10) = 17.66, Log likelihood statistics = 253.335, p 0.004; ρ = 0.690, Likelihood ratio test for ρ = 0 is 0.690, p = 0.027, Σ (Sigma) = 1.098; λ (Lambda) = 0.757; No of observation = 180, censored observations = 63, uncensored observations = 117.
Note: *,** and *** indicate significance at 10%, 5% and 1% levels, respectively; figures in the parentheses are probability values.

likelihood function (p = 0.004). It means that the coefficients of the explanatory variables used in the model are significantly different from zero. Also, the log likelihood ratio test rejected the hypothesis of the absence of correlation between the error terms of outcome (ε_1) and selection (ε_2) equations (ρ = 0.690, ρ = 0.027). This justifies the estimation of these two equations simultaneously using the Heckman selection model. Since the above equations were modeled using the maximum likelihood method, the coefficients of the explanatory variables do not represent their average impact on the dependent variable. So, we estimated the marginal impacts of the explanatory variables on the dependent variables, and these impact values are used to discuss the degree of influence of these variables on the dependent variables. The study shows that the impact of most of the explanatory variables is in line with their hypothesized direction. Also, the impact of some variables is different between the outcome and selection equations.

The age of HHH has a statistically significant positive impact on the volume of seed sold in the market. However, its impact on market participation is not significant. This finding is consistent with that of Omit et al. (2009).

There is a significant positive impact of cultivated land on seed sold volume in the market but its effect on market participation is not significant as in the case of the age of HHH. One ha increase in cultivated land leads to an increase of the seed volume sold by 6%. Irrigation also showed a significant positive impact on seed volume sold which would be increased by 4.2% with an increase in the irrigated land by 1%. These findings are also consistent with those of Azam et al. (2012).

In contrast to the above findings livestock, training and households' share in CBSPOs showed a significant positive impact on the households' participation in the market instead of volume of seed sold. As shown in Table 2, one unit increase in LSU leads to increase the probability of households' selling seed in the market by 7.1%. Similarly, there is a significant positive impact of training on market participation. Trained households' probability to sell rice seed in the market is 15.5% higher than the non-attendees. The better performance of trained households in market participation might be due to their superior skills on seed quality management and commercial orientation (Witcombe et al. 2010).

Households' share in CBSPOs also showed a significant positive impact on market participation. There is a 19% higher probability of selling rice seeds among households who have deposited shares in CBSPOs than their counterparts. The seed price shows a significant positive impact on households' decision to participate in the market. One unit increase in seed price (NRs kg^{-1}) increases the probability of households' selling seed in the market by 7.1%. Seed price has also an indirect impact on volume of seed sold in the market as shown from its marginal impact on seed sold volume (Table 2). One unit increase in seed price leads to an increase of seed sold volume by 11%. The other variables such as family labour, education of HHH and roof type did not show a significant impact on seed marketing but it does not mean that they do not have any role in households' decision in selling seed in the market and volume of seed sold.

Discussion

In this study, we found that about two-thirds of the farmers have participated in the rice market in Nepal, and that they sell a similar proportion of the total produced rice as seeds, that is, one-third of the farmers did not sell seed in the market. Previous studies have also noted the issue of the poor participation of farmers in the market in developing countries. Almekinders et al. (1994) and Wiggins and Cromwell (1995) found poor market participation of farmers, especially those facilitated by development projects, in seed marketing. In some cases, these projects failed to collect seed back which was provided to farmers as loans (Almekinders et al. 1994; Wiggins and Cromwell 1995). One of the reasons for poor performance of farmers in seed marketing was due to the poor focus of these projects in delivering technical and marketing skills. Our study also recognized the importance of training on rice seed marketing. We found that farmers had taken training on crop husbandry, seed quality maintenance and marketing from projects implemented by NGOs and government agencies. Though 78.3% of the households attended agricultural training, only 65% of the market partici-pants received marketing training from NGOs. It means some farmers got agricultural training from DADO (32% of the sample households) but the content of these trainings was focused on technical aspects of seed production with little or no information about marketing and entrepreneurship aspects. This might be due to the influence of the policy document called 'District Seed Self-Sufficiency Program' which is focused on technical aspects of seed production (Lal et al. 2009; Witcombe et al. 2010).

Another motivating factor for farmers to participate in seed marketing is the system of share holdings in CBSPOs. Only 50% of households reported that their CBSPOs have distributed profit generated from seed marketing to their members, but it has been mo-bilized as a loan to the members. About a quarter of the respondents (26%) have taken loans from their CBSPOs for household activities. In addition to providing loans to their members, the revenue collected at the organization level has been mobilized for the development of a seed processing facility (storage building and grading machine). The practice of share holdings by households is considered important from q social perspective as well because it enhances households' ownership towards their organizations (CBSPOs). Moreover, households with higher share contributions to CBSPOs hold more voting rights (value) in the decision making process. This norm is similar to that of private companies and not to the general cooperative principle where one member one vote is applied regardless of the distribution of share amount among the members (Acharya 2008). However, 75% of the households included in this study are organized either in informal groups or cooperatives.

Similarly, this study recognized that three additional economic variables—livestock holding, cultivated land and irrigation facility—have a significant positive impact on the rice seed selling behavior of farmers. The significant impact of livestock on households' participation in the market might be due to its contribution to soil fertility and thereby on crop yield. There is a significant positive correlation between LSU and crop yield ($r = 0.6$, $p = 0.02$). Livestock also contribute in households' cash income, but only 20% of farmers have received cash income from this sub-sector indicating the linkage of livestock on seed marketing is important mainly from its contribution towards increasing rice yield through improving soil fertility. Moreover, the impact of cultivated land and irrigation facility on seed selling volume is significantly different from zero. The correlation of these variables with rice yield is also positive, which implies that farmers with higher seed production area and better irrigation facility are more likely to sell higher quantities of seed in the market. However, strategies such as timely payment for seed or provision of credit and insurance system could address the small farmers' concerns, and those growing rice under rain-fed condition (Kugbei 2007).

Conclusions

Farmers' participation in rice seed production has been popular to increase access to diversified varieties in a cost effective way. However, to increase gains from seed production, farmers need to supply a maximum proportion of produced seed to the market. In this study, we analyzed the impact of households' socio-economic variables on the probability of households' rice selling seed in the market, and the volume of seed sold using a Heckman selection model. The result shows that 65% of farmers sell 64% of rice seed produced by their household. Out of the socio-economic variables, age of HHH, cultivated land and irrigation facility have major impacts on seed volume sold, whereas the major impact of livestock, training and share holding is on farmers' market participation. Seed price directly impacts on farmers' market participation, and has also an indirect impact on seed volume sold. So, from this study we found four important variables: training, irrigation, soil fertility and share holdings that can have policy implications in motivating farmers in rice seed marketing in the study area.

Endnotes

[a]SRR is the proportion of rice area covered by quality seed in the country in a year.

[b]LFU is a measure for labour force, where people from 15-59 years old regardless of their sex were categorized as 1 person = 1 LFU, but in case of children (10-14 years old) and elderly people (>59 years old) 1 person = 0.5 LFU.

[c]LSU is the aggregate of different types of livestock kept at household in standard units calculated using the following equivalents; 1 adult buffalo = 1 LSU, 1 immature

buffalo = 0.5 LSU, 1 cow = 0.8 LSU, 1 calf = 0.4 LSU, 1 pig = 0.3 LSU, 1sheep or goat = 0.2 LSU and 1 poultry or pigeon=0.1 LSU (CBS 2003).

[d]Earlier version of the paper was presented in the 11th international conference on dry land development: global climate change and its impacts on food and energy security in the dry lands, organized by International Dryland Development Commission, 18-23 March, 2013 at Beijing.

Abbreviations
CBS: Central Bureau of Statistics; CBSPOs: Community-based seed producers' organizations; CIMMYT: International wheat and maize research center; DADOs: District Agriculture Development Office; DISSPRO: District seed self-sufficiency program; FAO: Food and Agriculture Organization; FAOSTAT: Food and Agriculture Organization Statistics; HHH: Household head; MoAC: Ministry of Agriculture and Cooperative, Kathmandu, Nepal; SQCC: Seed Quality Control Center; SRR: Seed replacement rate.

Competing interests
The authors declare that they have no competing interests.

Authors' contributions
NPK: involved in different steps: research design, data collection, analysis and preparation of draft involved in the preparation of this article. KLM was involved in research design and editing of the document. Both authors read and approved the final manuscript.

Author's information
Narayan P Khanal is an Assistant Professor at the Graduate School for International Development and Cooperation (IDEC), Hiroshima University, Hiroshima, Japan.

Acknowledgements
We are thankful to Global Environmental Leader program of Hiroshima University for providing grant for field study. Similarly, supports from Forum for Rural Welfare and Agricultural Reform for Development (FORWARD Nepal) in data collection process.

References
Acharya BM (2008) Is cooperative really a democratic organization? Lessons learnt from the selected cooperative acts of the world: annual bulletin. Cooperative Development Board, Kathmandu, Nepal

Almekinders C, Louwaars N (1999) Farmers' seed production: new approaches and practices. Available via www.acss. ws/Upload/XML/Research/99.pdf. Accessed 12 April 2011

Almekinders C, Louwaars NP, Bruijin GH (1994) Local seed system and their importance for an improved seed supply in developing countries. J Euphytica 78:207–216

Azam MS, Ima KS, Gaiha R (2012) Agricultural supply response and small holder market participation: the case of Combodia. Available via http://www.rieb.kobe-u.ac.jp/academic/ra/dp/English/DP2012-09.pdf. Accessed 10 Jan 2013

Barrett CB, Marenya PP, Mcpeak JG, Minten B, Murithi FM, Oluoch-Kosura W, Place F, Randrianarisoa JC, Rasambainarivo J, Wangila J (2006) Welfare dynamics in rural Kenya and Madagascar. J Dev Stud 78(3):656–669

Benfica R, Tschirly D, Boughton D (2006) Interlinked transaction in cash cropping economies: the determinant of farmers' participation and performance in the Zambezi River Vally of Mozambique. Contributed paper presented at the 26th international association of agricultural economists conference, Gold Coast, Australia. August 12-18, 2006

Bishaw Z, Van Gastel AJG (2008) ICARDA's seed delivery approach in less favorable areas through village-based seed enterprises: conceptual and organizational issues. J New Seeds 9(1):68–88

CBS (2003) National sample census of agriculture Nepal, 2001/02: highlights. National Planning Commission Secretariat, Central Bureau of Statistics, Kathmandu, Nepal

David S (2004) Farmer seed enterprises: a sustainable approach to seed delivery? J Agric Hum Values 21:387–397

De Janvry A, Fafchamps M, Sadoulet E (1991) Peasant household behavior with missing market: some paradoxes explained. Econ J 101(409):1400–1417

FAO (2010) Promoting the growth and development of small seed enterprises for food security crops. Available via www.fao.org/docrep/013/i1839e/i1839e00.pdf. Accessed 10 March 2011

FAOSTAT (2012) Crop information 2011. Available via http://faostat.fao.org/site/567/DesktopDefault.aspx? PageID=567#ancor. Accessed 2 February 2013

Fafchamps M (1992) Cash crop production, food price volatility and rural market integration in the third world. Am J Agric Econ 74(1):90–99

Gauchan DM, Smale M, Chaudhary P (2005) Market-based incentive for conserving diversity on-farm: the case of rice landraces in central Tarai, Nepal. Genet Res Crop Evol 52:293–303

Heckman JJ (1979) Sample selection bias as a specification error. J Econometrica 47(1):53–161

Heltberg R, Tarp F (2002) Agricultural supply response and poverty in Mozambique. Food policy 27(1):103–124

Khanal NP, Maharjan KL (2013) Socio-economic determinants for the adoption of improved rice varieties in the tarai region of Nepal. J Int Dev Cooperation, Hiroshima University, Special Issue 19(4):17–27

Kugbei S (2007) Seed economics: commercial consideration for enterprise management in developing countries: international center for agricultural research in the dry areas. Scientific Publishers, India

Lal KK, Thapa M, Guenat D (2009) Review of seed projects in Nepal. Swiss Agency for Development and Cooperation Kathmandu, Nepal

MoAC (2010a) Statistical Information on Nepalese Agriculture. Ministry of Agriculture and Cooperatives, Singha Durbar, Annual Report, Nepal

MoAC (2010b) Seed production groups and cooperatives in Nepal. Ministry of Agriculture and Cooperative, Nepal

Nawata K (2004) Estimation of the female labor supply model by Heckman's two-step estimator and maximum likelihood estimator. Math Comput Simul 64:385–392

Nawata K, Nagase N (1996) Estimation of sample selection method. Econometric Rev 15:387–400

Omiti JM, Otieno DJ, Nyanamba TO, Mccullough E (2009) Factors influencing the intensity of market participation by smallholder farmers: a case study of rural and peri-urban areas of Kenya. Available via http://ageconsearch.umn.edu/bitstream/56958/2/0301Omiti%20-%20FINAL.pdf. Accessed 4 December 2012

Pindyck RS, Rubinfield D (1981) Econometric models and economic forecasts. McGraw Hill, New York, USA

Pokhrel S (2012) Role of DISSPRO and CBSP in current seed supply situation in Nepal. J Agric Environ 13:53–59

Setimela PS, Monyo E, Banjiger M (ed) (2004) Successful community based seed production strategies. CIMMYT, Mexico

Srinivas T, Zewdie B, Javed R, Abdoul A, Rahaman A, Amegbeto K (2010) ICARDA's approach in seed delivery: technical performance and sustainability of village-based seed enterprises in Afghanistan. J New Seeds 11(2):138–163

SQCC (2011) Seed balance sheet 2011. Available via http://sqcc.gov.np/. Accessed 4 February 2012

Wiggins S, Cromwell E (1995) NGOs and seed provision to smallholders in developing countries. World Dev 23:413–422

Witcombe JR, Devkota KP, Joshi KD (2010) Linking community-based seed producers to markets for a sustainable seed supply system. J Exp Agric 46(4):425–437

5

Alternative food chains as a way to embed mountain agriculture in the urban market: the case of Trentino

Emanuele Blasi, Clara Cicatiello[*], Barbara Pancino and Silvio Franco

* Correspondence:
cicatiello@unitus.it
Department of Economics and
Management, Università degli Studi
della Tuscia, via del Paradiso 47,
Viterbo 01100, Italy

Abstract

Peri-urban agriculture is exposed to multiple pressures, which push to diversification and multifunctionality. However, the urban-rural link develops in different ways according to the features of the territories. A very interesting case is that of mountain areas, where the proportion among urban and rural domain is very skewed towards the latter and, at the same time, farms face major environmental problems that in most cases jeopardize their competitiveness in the mainstream market. Alternative food chains may play a key role in these contexts, especially for their ability to put farms in touch with the demand of consumers living in the urban areas located in the valley. In this paper we study the case of Trentino, an Italian Alpine region where alternative food chains are quickly developing, by comparing the development of alternative markets in this context with other Italian peri-urban areas. The mountain environment makes it very difficult for farms to standardize their products according to the requirement of the large retailers. Through alternative food chains, the typicality of products and the savoir faire of the farmers – representing the two main factors of products' added value – are endorsed and more easily communicated to the market. Data from a survey conducted on short food chain consumers show that they are inherently more careful to these particular cues of the products, as a result of a lifestyle that makes them more attached to identity and origins, as well as being more proud of their territory. These evidences confirm that in the Trentino area, for its structural and cultural traits, alternative food markets are meaningful for the survival and development of the local agricultural sector.

Keywords: Alterative food networks; Short food supply chains; Peri-urban agriculture; Rural-urban link; Mountain farming; Local products

Background

Alternative food networks and urban proximity

Recent research on the role of agriculture in urban and peri-urban contexts raises many questions about how the urban–rural link develops in regions and countries with different characteristics. Indeed, the problem of planning sustainable food production, distribution and consumption patterns should be related to the specificities of different territorial contexts. In this paper we investigate the emergence of local food chains in different contexts, focusing on the case of mountain areas, where these practices may play an important role in fostering the maintenance and the sustainability of agriculture.

The development of Alternative Food Networks (AFNs; Renting et al., 2003) is a central aspect of the urban–rural link, as they are very likely to develop in peri-urban contexts where the traditional agricultural functions are often replaced by non-agricultural or post-productive ones (Luttik and van der Ploeg, 2004). Defining AFNs is difficult, as the term has been used in different contexts and for different purposes. What most of the definitions share is an element of sense-making against the market and the rationale of agricultural industrialization (Holloway et al., 2006). It follows that AFNs are usually associated with all those local and ethical food chain systems which differ from mainstream food supply systems (Holloway and Kneafsey, 2004). Four main characteristics of AFNs are usually recognised: (1) a short distance between producers and consumers; (2) the involvement of small farms with a preference for ethical and responsible modes of production, in contrast with the industrial agribusiness approach; (3) the existence of food purchasing venues (both material and intangible) such as food cooperatives, farmers' markets, websites and so on; (4) a commitment to the social, economic and environmental dimensions of sustainable food consumption, distribution and production (Jarosz, 2008).

In the research on AFNs, many studies focus on how short food chain practices develop under different conditions. Short food supply chains (SFSCs) have been central to recent research on the emergence of alternative forms of agriculture and food supply in Western countries, as they are often understood as a consequence of the so-called turn to quality expressed by consumers in the food domain (Goodman, 2003). Indeed, in the mind of consumers they are associated with more traditional, locally embedded and sustainable farming practices (Ilbery and Maye, 2005), although this perception may sometimes be incorrect, as "local" is not in itself a guarantee of "a strong turn to quality based production" (Winter, 2003).

Rooted in the context of the emergence of many alternative forms of agriculture and food networks, SFSCs are characterized by the proximity of producers and consumers, either relational – i.e. with few (or even no) intermediaries between producers and consumers – or physical – i.e. a short geographical distance between them (Aubry and Kebir, 2013). Ideally, they fulfil both these conditions (Pascucci, 2010). These features are inherent to different types of SFSCs, ranging from direct sale at farm shops to farmers' markets, internet sale and organized box schemes. The considerable social involvement of the participants is another key feature of SFSC practices, due to both their embeddedness in the local community (Sage, 2003) and the social interactions that are likely to take place in these alternative markets (Cicatiello et al., 2014).

How does the urban–rural link contribute to the emergence of SFSC practices? The proximity to urban centres provides farmers with both an opportunity and an incentive to restructure their operations with a multifunctional approach. Namely, the development of alternative food chains is highly influenced by two main processes: urbanisation and rural restructuring, together with their mutual interaction (Jarosz, 2008). In past decades, diversification had been observed as a survival strategy in rural areas characterised by urban pressure (Ilbery, 1987; Bryant and Johnston, 1992), where consumer demand for agricultural goods and services appears to be much stronger than elsewhere.

First, agriculture is needed to help in the preservation of a quality environment and landscape. Indeed, agriculture plays a key role in managing the peri-urban landscape and the social, aesthetic and environmental functions of urban agglomerations nearby

(Davoudi and Stead, 2007). Nonetheless, food production remains an important function of agricultural activities located in peri-urban areas. It has been observed that consumers increasingly prefer local produce because in local products they recognize added value related to tradition, quality and naturalness. As a result, consumers are increasingly willing to buy directly from producers, through alternative food chains at the local level (Pearson et al., 2011).

In response to these requests from urban consumers, more and more farms located in the surroundings of towns and cities are trying to meet this increasing demand, by offering recreational/leisure activities and making their products available to a greater share of urban consumers. In practice, this involves enhancing environmentally-friendly farming procedures, hobby farming, recreation-oriented diversification, social farming, short food supply chains and direct marketing (Zasada, 2011).

Of course, how farms respond to consumers' evolving demand is highly influenced by the territorial context where these processes take place. Namely, the geographical constraints of peri-urban areas, the features of the local agriculture (which may be more or less suited to multifunctional activities), the culture and traditions typical of each place, are all key factors to understanding the urban–rural link in different locations. In this paper we focus on the emergence of local food chains in a peri-urban mountain context, taking the Italian alpine region of Trentino as the focus of a case study. The particular interest of this case lies in the characteristics of mountain areas where, on the one hand, the traits of the pre-existing agricultural activities might encourage the development of multifunctional activities and, on the other hand, the link with the urban domain is perhaps weaker than elsewhere for geographical reasons. We argue that these factors are able to shape the development of local food chains by influencing consumer and producer approaches to alternative markets, so as to distinguish them from the rural/urban interactions that take place on the urban fringe of the cities. In order to explore these issues in the following sections we report the results of an empirical study performed in 2012 on 745 SFSC consumers, both in Trentino and in four peri-urban contexts. The exploratory study conducted to understand the development and features of local food chains allows us to identify the unique elements of SFSC development in Trentino through a comparison with SFSC practices in other Italian peri-urban areas. This comparison will highlight how studying the specificities of such a mountain context may improve understanding of how local specificities and constraints affect or enhance the development of short food chains, thus contributing to the ongoing research into these practices.

Investigating SFSCs in Trentino
Context
Mountain areas represent a very interesting context for studying the development of alternative marketing networks. Indeed, in these areas the proportion of urban and rural domains is very skewed towards the latter and, at the same time, farms have to work in a hard environment, which in most cases jeopardizes their competitiveness in the mainstream market. This is the reason why alternative food chains may play a key role to the survival of agricultural operations in these contexts, especially thanks to their ability to put farms in touch with the demand of consumers living in the urban areas located in the valley, thus safeguarding their survival.

In Trentino SFSCs are well developed, with about 10% of farms selling their products directly to consumers, mostly by means of farmers' markets and direct sale at the farm as well as with more innovative means, such as solidarity purchase groups and community-supported agriculture (Marino et al., 2013). Many highly-innovative practices and experiments have been reported there. This is due to the particular features of the local agricultural sector which is made up almost exclusively of small and micro-farms, often located in areas with poor accessibility and facing environmental, climatic, geographic and logistical challenges which affect their competitiveness on the market. These farms play a very important social and environmental role in representing the outpost of human influence in amidst the mountain wildlife, and their survival has so far mainly depended on widespread cooperation. This type of supply chain, while it guarantees the farmer – whether professional or part-time - a reliable and fairly profitable market, it certainly limits independence and decision-making power over which strategies to implement (Raffaelli et al., 2009).

This is perhaps the reason why some farms have started to experiment with new, more direct and autonomous ways of putting their products on the market, with the aim of maximising the value of their specificities. Alpine huts were the first farms to develop in this sense: each estate produces its own cheese, which is unique as its qualities are determined by a set of geographic variables (altitude, exposure of the slope), inputs (such as grazing) and technical skills (know-how of the farmer). The cheese is sold directly on site to tourists who come during the summer or in the villages, when the farmer goes back to the valley at the beginning of the autumn.

Practices such as these have spread more widely in recent years, and with the opening of the first farmers' markets, many other farms have entered SFSCs. Coldiretti, the major Italian farmers association, has acted as the main promoter of the diffusion of farmers' markets, in this region as well as in other parts of Italy. In 2006 it promoted the first farmers' market in Trentino, located in the central square of Trento, which was then followed by 10 other markets organized every week in the main urban centers of the region. In these markets locally-produced cured meats, cheeses and numerous heritage varieties of fruit and vegetables from the different valleys are sold, their value maximised through direct contact with consumers.

Beyond farmers' markets and direct sale on the farms, the region has also launched innovative forms of SFSCs. Solidarity purchase groups (henceforth referred to as GASs using the Italian acronym Gruppo di Acquisto Solidale) are widely spread in the area, where they have developed following a well-established trend in many Italian regions (Cembalo et al., 2013). As many as 16 of these practices may be counted in the city of Trento alone, while at least another 14 operate in different towns in the region. A GAS often emerges from informal associations based on the sharing of common values, such as ethical consumerism, attention to the environment, proximity among families, the desire to support local agriculture and so on (Migliore et al., 2012). Some early experiences of Community Supported Agriculture are also reported, such as the "Adopt a Cow" scheme, which involves consumers in the sustenance of cattle in winter, thus providing them with dairy products to be collected at the Alpine huts during the summer.

However, the different types of SFSC operating in the Trentino region seem much less urban-centric than in other areas studied. Indeed, the city of Trento, although it counts over 100,000 inhabitants and is surrounded by agricultural land, appears to be quite distant – in both geographical and relational – from the rural domain that in this region is primarily located in the mountains. This is primarily because the mountainous terrain in the region makes communication difficult, but also because there are so many small but very cohesive and distinct local communities.

Methods

A survey on the development of SFSCs was carried out in the Trentino region in 2012. It was part of a larger project funded by the Italian Ministry of Agriculture and carried out by CURSA (University Consortium for Socioeconomic and Environmental Research), with the aim of analysing the features and development of different types of SFSCs in a variety of territorial contexts. To do this, consumers participating in different forms of SFSCs were surveyed in 5 Italian areas (two metropolitan areas, two middle-size towns and one mountain area – Trentino). For the purpose of this paper, the data gathered through this survey made it possible to analyse the features of SFSCs in Trentino in comparison with the similar practices in other Italian peri-urban contexts, so as to highlight the specificities of the area.

Given the exploratory aim of the study, a mixed quantitative-qualitative research approach was applied to the survey, by administering a questionnaire directly to the consumers and conducting interviews with farmers. We used convenience sampling, which is well-suited to exploratory studies, as it provides a gross estimate of the result, although it remains a non-probabilistic sampling method (Guerrero et al., 2010).

The study included a total of 39 SFSC practices, among which:

- 16 farm shops (2 in Trentino);
- 11 farmers' markets (1 in Trentino);
- 9 GASs (2 in Trentino)
- 3 CSA practices (1 in Trentino – the above-mentioned Adopt a Cow scheme).

The questionnaire was made up of 9 closed questions, concerning:

- general consumer food shopping habits (food expenditure, habitual food stores etc.);
- consumer food shopping habits at SFSCs (expenditure, frequency of participation, products purchased etc.);
- consumer motivations for participating in SFSCs;
- information about consumer demographic profiles.

The questionnaire was administered to consumers shopping at farm shops, farmers' markets, GASs and CSA experiences. In the first two cases, consumers were interviewed face-to-face directly at the shops or the markets, after having completed their purchases. Instead, as GAS and CSA consumers usually keep in contact with the organizers through emails, in these cases the questionnaire was administered in an online version and sent to their email address. The data retrieved were all fed into an online structure so as to build a single database that was used as the basis for data processing. In total, 745 SFSC consumers (142 of which in Trentino) participated in the survey.

For the purpose of this paper, we will analyse the results of the survey by comparing the results obtained in Trentino with those related to SFSCs in other areas, with the aim of highlighting the features that SFSC practices assume in the particular context of Trentino. The differences between Trentino and the other contexts were tested using ANOVA and Chi-square analyses so their statistical significance could be considered in the discussion.

We also performed in-depth interviews with 12 farmers involved in the SFSC practices analysed in the survey held in Trentino, in order to get producers' opinions about their operations and development. The interviews were carried out face-to-face, directly at the farms or, for the 4 farmers involved in the farmers' market, at the end of the market. The interviews were carried out as a conversation with the farmers, using the following outline:

- main features of the farm;
- marketing channels of the farm;
- motivations for joining SFSC practices;
- pros and cons of participation in SFSCs.

The qualitative information provided by the farmers during the interviews is explored in the discussion to enrich the analysis and more deeply understand the specific features of SFSC practices in Trentino.

Results and discussion

Out of 745 respondents to the questionnaire directed at SFSC consumers, 142 consumers were interviewed in Trentino. They are divided among the different chain types as shown in Table 1.

In about 60% of cases the respondent was a woman. Both in Trentino and in the other territorial contexts a higher quota of women was reported among the respondents interviewed at farmers' markets and farm shops than among GAS consumers. The average age of the consumers was about 50, which may explain the small size of the average household and the absence of children among the components of the households (on average less than 1 child per family). However, the respondents interviewed in Trentino were a little younger, although this difference was not very significant and their families were larger: 14% were made up of 5 or more components whilst in the peri-urban contexts analysed large families represented less than 7% of the sample.

Table 1 Respondents divided per SFSC type in the different contexts analysed

SFSC type	Trentino		Peri-urban contexts	
	Number of respondents	*Quota of respondents*	*Number of respondents*	*Quota of respondents*
Farmers' market	31	22%	258	43%
Farm shop	34	24%	209	35%
GAS	71	50%	115	19%
CSA	6	4%	21	3%
Total	142	100%	603	100%

Despite these findings, in the households interviewed in Trentino the weekly expenditure for food was lower than in the peri-urban areas covered by the survey (€98 in Trentino vs. €111 elsewhere, p-value = 0.013).

We asked consumers to state how much they usually spend when they purchase at SFSCs; results show that the average expenditure is about €27. Consumers from Trentino spend a little more for every purchase withSFSCs than the other respondents (p-value = 0.0004), although this finding is not consistent across the SFSC types: higher expenditures are recorded for the households participating in GASs, probably as a consequence of the concentration of purchases in the orders, which are typically completed only once a month. Instead, farmers' markets and farm shop customers typically make their purchases once a week.

Most of the consumers interviewed had been purchasing at SFSCs for a long time: nearly half of them had been purchasing at SFSCs for more than 2 years, while only 5% were interviewed during their first contact with these practices. This finding is even more pronounced in Trentino than elsewhere, and chi-square analysis (p-value = 0.085) confirms that Trentino consumers are more likely to have been purchasing at SFSC for more than 2 years than the other respondents (Figure 1). These results support the idea that Trentino is a stronghold of SFSC practices, with a loyal consumer base. Indeed, it seems that the recent development of SFSCs in this area is rooted in a strong attraction to alternative markets by consumers, which likely already existed.

How do SFSC consumers find out about these alternative opportunities for food purchase? As many as 48% of consumers surveyed mentioned "word of mouth" as the main source of information about SFSCs. In the case of farmers' markets and farm shops, physical proximity also plays a role, since these experiences provide a permanent location which is easy for consumers to notice when they pass nearby. In Trentino only 9 consumers (out of 142) mentioned formal communication sources such as internet and advertisements, whilst 60% stated they got to know about SFSCs experiences through word of mouth. Compared to the consumers interviewed in peri-urban contexts, they seem to value informal communication even more, as is shown in Figure 2 and confirmed by a chi-square analysis (p-value = 0.013).

Within the territorial context of Trentino, perhaps more than elsewhere, the communication strategies often used for marketing purposes are not suitable for SFSCs, whose

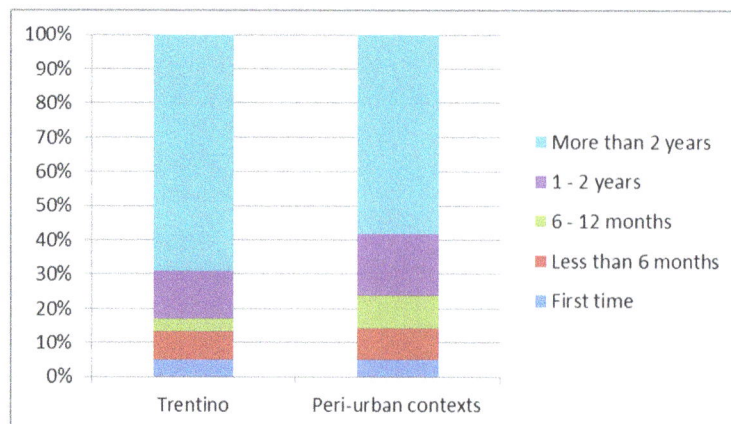

Figure 1 Results for the question "How long have you been purchasing with this SFSC?"

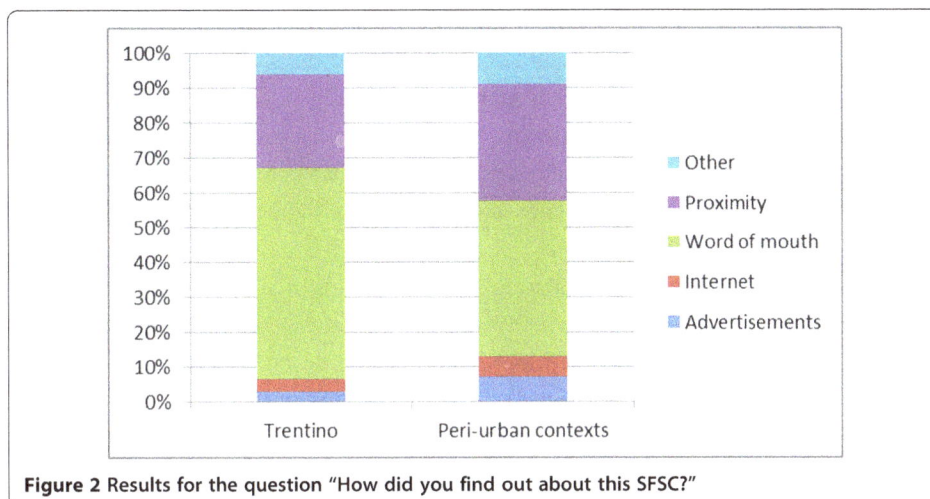

Figure 2 Results for the question "How did you find out about this SFSC?"

selling points can be communicated more effectively through direct contact between people rather than through the media. This is probably due to the very strong social network that still binds communities in the mountainous areas. The small size of villages, the sense of solidarity among people living in the mountains and the institutional effort to keep the local communities active result in a tendency to share ideas, opinions and events with neighbors and fellow citizens. This is likely to help spread information through informal channels much more than happens in peri-urban areas, thus making advertisement and other conventional communication strategies less effective and almost unnecessary. On the contrary, physical proximity assumes quite an import role in peri-urban contexts, where the wider adoption of SFSC practices in the urban network enhances the possibility for consumers to get acquainted with these practices in the course of their daily errands.

Interesting issues emerge from the analysis of consumer behavior at SFSCs. First, the products consumers buy with greatest frequency vary according to the type of SFSC. Almost all consumers buy fruit, vegetables and other fresh products at farm shops and farmers' markets, whereas GAS consumers are more likely to purchase processed non-perishable products, probably due to the longer time gap between deliveries. This is true in Trentino as well as in the other contexts analysed.

However, as we look at the marketing channels where consumers usually shop for food, some interesting differences emerge. In the questionnaire consumers were asked to indicate the quota of their total food purchases they usually buy through SFSCs and how the remaining quota is divided among other marketing channels. Their responses and statistical analyses confirm that in general consumers from Trentino have the tendency to concentrate their purchases at supermarkets, whereas urban consumers - although they still prefer this marketing channel - display higher diversification in their choice of food outlets. Urban consumers probably have greater access to other types of food distributors such as grocery stores, discount retailers and large hypermarkets with respect to consumers living in mountainous areas, who for geographical reasons (distance to the nearest village/town, poor roads, snow in winter and so on) find it more difficult to travel. Therefore, when they do, they are likely to go to a single food store to stock up for a week or even longer. Beyond this somewhat expected finding, it's very interesting to look at the quota of their food consumers purchase at SFSCs. Here, a remarkable difference can be found across the SFSC types and between Trentino and peri-urban contexts, which clearly emerges by looking at the data presented in Figure 3, although this finding lacks statistical significance given the small sample size. GAS participants are much more devoted to SFSCs than other consumers, as they buy over 35% of their food through GASs. This statistic is a little higher for Trentino, possibly because many consumers live outside the city in quite large houses, where they can stock large amounts of products. Farmers' markets and farm shops customers buy about 25% of their food at SFSCs, although for the farmers' market customers interviewed in Trentino this number is much lower than in peri-urban contexts. This may be due to the weekly opening of the farmers' markets operating in this area, whereas in the other contexts many farmers' markets open more often: twice a week or in larger cities such as Rome and Turin, even every day. Finally, as concerns CSA experiences, people participating in the Adopt a Cow project are much less devoted to alternative food provisioning channels, showing they are "newcomers" to SFSCs. On the contrary, the other CSA practices examined in the survey elsewhere provide participants with a higher quota of their food provisions.

Therefore, different forms of SFSCs have different capacities to act as major food provisioning channels for consumers: while a GAS is likely to become a dominant choice for food purchasing, other forms of SFSCs seem to be more suitable for complementing mainstream marketing channels. Namely, in Trentino more than elsewhere, the choice to participate in a GAS practice represents a crucial decision in the management of the food purchases for a household.

Finally, we analyzed the motivations that drive consumers to SFSCs. In general, the most important reasons refer to the distinctive elements of the products that can be purchased in these alternative chains: quality, healthfulness, local origin. GAS consumers mention ethical aspects such as environmental concerns and trust in the farmers more frequently, whereas farmers' markets and farm shops consumers have a weaker ethical set of motivations and mainly focus on the individual benefits that SFSC may deliver (e.g. quality and healthful products). At the other end of the ranking we surprisingly find motivations related to price and savings, which are scored much lower than the others. Although similar results have already emerged from

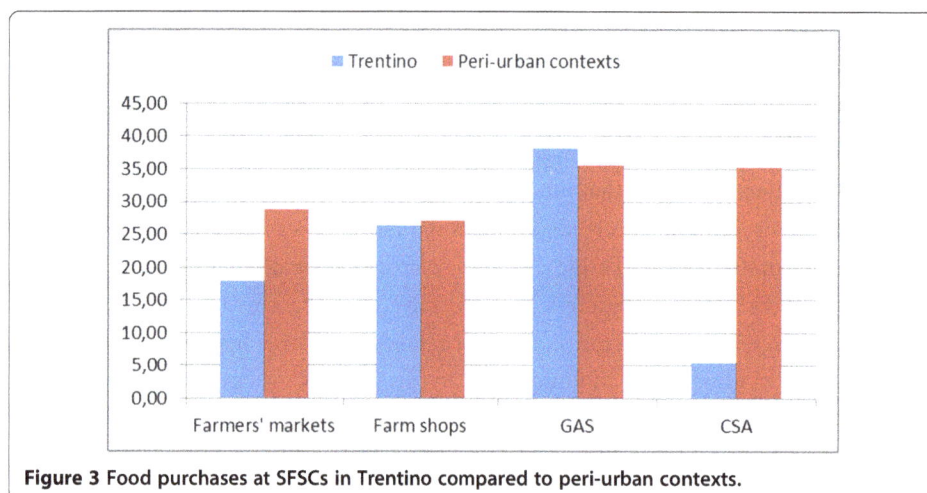

Figure 3 Food purchases at SFSCs in Trentino compared to peri-urban contexts.

similar research done on Italian farmers' markets (Pascucci et al., 2011), these findings confirm that the search for the lowest price is not a key driver for consumers using all types of SFSCs.

However, some differences can be found among the different territorial contexts, as it is shown in Table 2. Namely, consumers from Trentino seem to be more concerned about their health and the environmental impact of their purchases than urban consumers; at the same time, although saving money is one of the less important motivations in general, consumers interviewed in Trentino do value this aspect a little more than the others. It seems that, on the one hand, their decision to participate in SFSCs is influenced by their belief that the products have a greater environmental value but, on the other hand, they still keep an eye on the economic aspect of the purchase. Instead, for urban consumers the convenience of alternative markets, i.e. their proximity, accessibility etc., is more important than for consumers living in a mountain context, although these differences are not statistically significant.

Greater insight into the specificities of SFSCs in Trentino emerges from the in-depth interviews we conducted with 12 producers participating in these experiences. Among them, 4 participated in a farmers' market, 3 belonged to a GAS, 3 sold their products at the farm and 2 were involved in the "Adopt a cow" scheme. Their farms were quite diverse both in size, ranging from very small (less than 1 hectare) to well over 50 hectares, and in location, with 5 farms located very high in the mountains, above 1,000 metres. This is maybe the reason why their points of view regarding SFSCs and the market in general is varied and diverse. Typical production at these farms included dairy products as well as fruit and vegetables, in line with the area's specialisation in livestock, apples and soft fruit.

Concerning their relationship with SFSCs practices, all the farmers said that SFSCs – very often more than one, as they tend to participate in several – is an important marketing channel, which absorbs a remarkable quota – often most of – their total production. However, some of them are still linked to the local farmers' cooperatives, which represent a low-risk option for selling their products. This tendency to specialize in SFSCs is the consequence of a general appreciation of these practices by the farmers, who state they are able to find, thanks to these networks, a marketing niche at the local level.

Among the reasons that support such a broad appreciation of the short chain, most of the farmers mention the desire to let consumers know more about the farm as well as a wish to have a direct relationship with them. This approach to SFSCs seems to go hand-in-hand with the trust consumers show towards the farmers, which may be the result of a positive exchange between them during purchases. The economic motivations, such as selling at a good price and retaining more value from the sale of quality products, are also quite important to farmers participating in farmers' markets or selling their

Table 2 Motivations of surveyed consumers for purchasing at SFSCs

Contexts	Motivations						
	Healthful products	Quality products	Local products	Environmental concerns	Save money	Convenience	Trust in producers
Trentino	4.85*	4.80	4.53	4.63*	3.50**	3.23	4.36
Peri-urban contexts	4.75*	4.72	4.55	4.48*	3.23**	3.41	4.40

**The difference between Trentino and the other contexts is significant for α < 0.05.
*The difference between Trentino and the other contexts is significant for α < 0.10.

products at a farm shop. Clearly, the price applied at these SFSCs - although not a significant driver to attract customers according to our survey, clearly satisfy the producers, at least when they value the earnings in comparison with mainstream marketing channels.

A primary motivation among farmers participating in GASs was their desire to be appreciated for the quality of their products, while the economic implications were much less important, although they said they were satisfied with the extra income they made through SFSCs. Hence, they have a very different approach to SFSCs, which matches that of GAS consumers.

The approach of farmers participating in the Adopt a Cow scheme was again different. They work under high environmental and logistic constraints, managing pastures located at considerable altitudes (1500–1700 m). Their participation in the project is therefore understood as a way to improve people's awareness of their traditional way of farming, thus improving tourism in the area, with significant economic implications, especially for those estates that have hospitality facilities available.

Therefore, it seems that producers' views of SFSCs match those of their consumers in several ways. In Trentino SFSC practices, perhaps more than elsewhere, the mutual influence of SFSC participants seems to shape the specific features of these practices in the particular context where they develop.

Conclusions

The conservation of agriculture in peri-urban environments is able to deliver environmental and recreational value to the city, as well as supporting the survival of local farms. In peri-urban contexts the role of traditional agricultural is likely to be replaced by non-agricultural or post-productive alternatives, thus pushing farms to move towards a multifunctional approach. These processes often support the development of AFN practices that complement rural/urban links.

In this paper we have analysed how such innovative food networks are shaped by the specific context of the Italian alpine region of Trentino.

Our analysis revealed that the Trentino area, due to its structural and cultural traits, is very well-suited for developing these practices, despite its geographical and logistical problems caused by its mountainous location. However, as a consequence of these specific traits, SFSCs have developed quite differently from in other Italian peri-urban contexts. Namely, by analysing the responses of SFSC producers and consumers in interviews and a survey in 2012, some interesting issues have emerged. For example, the consumers interviewed appear to be more loyal to SFSC practices than elsewhere, although they have emerged quite recently. Informal means of communication have proven to be very important to this purpose. At the same time, as expected, the geographical and logistical constraints typical of a mountain area seem to influence approaches to alternative markets, which are not likely to become a main source of food provision unless the SFSC scheme brings the products directly to consumers' homes, as is in the case of GASs. This is remarkably different from more urban-integrated SFSC practices in which the SFSCs are more accessible and integrated into the everyday life of consumers, just like mainstream food stores.

By comparing the results of the survey administered to consumers with the opinions of farmers we talked to during in-depth interviews, it seems they share a similar approach to SFSC practices, probably as a consequence of the strong social capital they share, typical of mountain areas and small communities. This is maybe the reason why motivations related to quality and relations with other members of the community are key drivers for both to join SFSC practices.

In the end, SFSCs seem to be able, in the Trentino context, to reinforce the link between the mountains and the city. Indeed, it seems that consumer attitudes toward these practices are already very positive, thanks to their strong territorial identity. It follows that, rather than the farms responding to the needs of consumers as mostly happens elsewhere, in Trentino it is the consumers who satisfy the needs of the farms, for the sake of this bond. As we have seen, what farms need is mainly to protect their specific features and to communicate them to the market. Indeed, most of these farms could not be economically sustainable if they operated on the mainstream market, thus one of the few options they have is to take advantage of the uniqueness of their products, which are special both in terms of tangible (taste, traditional recipes, etc.) and intangible qualities (know-how of the farmer, relation to the local identity, etc.) to carve out a niche in the market. On the other hand, these consumers are inherently more aware of these needs, as a result of a lifestyle that makes them more attached to their identity and origins, as well as proud of their territory and the products that it can provide.

Therefore, there seems to be a natural match between the need of the farmers and the new demands by consumers. It follows that, despite the small size of the local market - only 3 of the 217 municipalities in Trentino have more than 20,000 residents – SFSCs have developed in a number of forms, thus contributing to the sustainability of the local agricultural sector. Of course, before considering the findings of this study for local policy planning, further research should be performed in order to confirm the characteristics observed. However, it seems that in the Trentino area SFSCs have the potential to play a key role in driving the development – or at least supporting – the agricultural sector in the area, and, at the same time, to foster social cohesion between rural and urban communities. From this perspective, it would be appropriate to support these practices in mountain communities, through rural development programs and local support of SFSC practices.

Of course, these findings need to be confirmed in further studies of other mountain communities. These might focus on the extent to which, in other international contexts, the same differences between mountain and peri-urban SFSCs may be found. Indeed, there is a gap in the research regarding how short food chain practices develop in particular environments such as the mountains. In-depth research in this domain may help to identify the opportunities that the development of SFSCs are likely to offer in these contexts. How can SFSCs contribute to supporting mountain agriculture and its sustainability? At the policy level, is it appropriate to support these farmsas a means to improve food system sustainability in mountain communities? Can the strengthening of local identity (e.g. through protected designations of origin) foster the development of SFSCs? These and other related questions provide the basis for an interesting stream of research to be undertaken in the coming years.

Abbreviations
AFN: Alternative food networks; GAS: Solidarity Purchase Group (from the Italian acronym Gruppo di Acquisto Solidale); SFSC: Short food supply chain.

Competing interests
The authors declare that they have no competing interest.

Authors' contributions

EB wrote the Results and discussion section. CC wrote the Background section. BP wrote the Method section. SF wrote Conclusions. All authors read and approved the final manuscript.

Acknowledgements

The authors wish to thank the University Consortium for Socioeconomic and Environmental Research (CURSA) and the coordinator of the project about short food supply chains in Italy, some data of which are here analysed. Our acknowledgements go to the researchers who worked on the project by carrying out the survey all over Italy. We also thank the anonymous reviewers whose comments were very useful to improve the structure and content of the paper.

References

Aubry C, Kebir L (2013) Shortening food supply chains: a means for maintaining agriculture close to urban areas? the case of the French metropolitan area of Paris. Food Policy 41:85–93

Bryant CR, Johnston TRR (1992) Agriculture in the City's Countryside. Belhaven Press, London

Cembalo L, Migliore G, Schifani G (2013) Sustainability and new models of consumption: the Solidarity Purchasing Groups in Sicily. J Agric Environ Ethics 26(1):281–303

Cicatiello C, Pancino B, Pascucci S, Franco S (2014) Relationship patterns in food purchase: observing social interactions in different shopping environments. J Agric Environ Ethics 27 (4). doi:10.1007/s10806-014-9516-9

Davoudi S, Stead D (2007) Urban–rural-relationships: an introduction and brief history. Build Environ 28:269–277

Goodman D (2003) The quality 'turn' and alternative food practices: reflections and agenda. J Rural Stud 19(1):1–7

Guerrero L, Claret A, Verbeke W, Enderli G, Zakowska-Biemans S, Vanhonacker F, Issanchou S, Sajdakowska M, Signe Granli B, Scalvedi L, Contel M, Hersleth M (2010) Perception of traditional food products in six European regions using free word association. Food Qual Preference 21(2):225–233

Holloway L, Cox R, Venn L, Kneafsey M, Dowler E, Tuomainen H (2006) Managing sustainable farmed landscape through 'alternative' food networks: a case study from Italy. Geogr J 172(3):219–229

Holloway L, Kneafsey M (2004) Producing-consuming food: closeness, connectedness and rurality. In: Holloway L, Kneafsey M (eds) Geographies of Rural Cultures and Societies. Ashgate, Aldershot, pp 262–282

Ilbery B (1987) The development of farm diversification in the UK: evidence from Birmingham's urban fringe. J Royal Agric Soc England 148:21–35

Ilbery B, Maye D (2005) Food supply chains and sustainability: evidence from specialist food producers in the Scottish/English borders. Land Use Policy 22:331–344

Jarosz L (2008) The city in the country: growing alternative food networks in metropolitan areas. J Rural Stud 24(3):231–244

Luttik J, van der Ploeg B (2004) Functions of agriculture in urban society in the Netherlands. In: Brouwer F (ed) Sustaining Agriculture and the Rural Economy: Governance, Policy and Multifunctionality. Edward Elgar, Cheltenham, pp 204–222

Marino D, Cavallo A, Galli F, Cicatiello C, Borri I, Borsotto P, Di Gregorio D, Mastronardi L (2013) Esperienze di filiera corta in contesti urbani. Alcuni casi studio. Agriregionieuropa 32(9):1–7

Migliore G, Cembalo L, Caracciolo F, Schifani G (2012) Organic consumption and consumer participation in food community networks. New Medit (Suppl) 11(4):46–48

Pascucci S (2010) Governance structure, perception, and innovation in credence food transactions: the role of food community networks. Int J Food Syst Dyn 1(3):224–236

Pascucci S, Cicatiello C, Franco S, Pancino B, Marino D (2011) Back to the future? understanding change in food habits of farmers' market customers. Int Food Agribus Manag Rev 14(4):105–126

Pearson D, Henryks J, Trott A, Jones P, Parker G, Dumaresq D, Dyball R (2011) Local food: understanding consumer motivations in innovative retail formats. Br Food J 113(7):886–899

Raffaelli R, Coser L, Gios G (2009) Esperienze di filiera corta nell'agro-alimentare: un'indagine esplorativa in provincia di Trento. Econ Agro-alimentare 1:25–42

Renting H, Marsden TK, Banks J (2003) Understanding alternative food networks: exploring the role of short food supply chains in rural development. Environ Plan A 35(3):393–412

Sage C (2003) Social embeddedness and relations of regard: alternative 'good food' networks in south-west Ireland. J Rural Stud 19(1):47–60

Winter M (2003) Embeddedness, the new food economy and defensive localism. J Rural Stud 19(1):23–32

Zasada I (2011) Multifunctional peri-urban agriculture. a review of societal demands and the provision of goods and services by farming. Land Use Policy 28(4):639–648

6

Oil and food prices in Malaysia: a nonlinear ARDL analysis

Mansor H Ibrahim

Correspondence:
mansorhi@hotmail.com
International Center for Education
in Islamic Finance (INCEIF), Lorong
Universiti A, 59100 Kuala Lumpur,
Malaysia

Abstract

The present paper analyses the relations between food and oil prices for Malaysia using a nonlinear autoregressive distributed lags (NARDL) model. The bounds test of the NARDL specification suggests the presence of cointegration among the variables, which include the food price, oil price and real GDP. The estimated NARDL model affirms the presence of asymmetries in the food price behavior. Namely, in the long run, we find a significant relation between oil price increases and food price. Meanwhile, the long run relation between oil price reduction and the food price is absent. Furthermore, in the short run, only changes in the positive oil price exert significant influences on the food price inflation. With the absence of significant influence of oil price reduction on the food price both in the long run and in the short run, the role of market power in shaping the behavior of Malaysia's food price is likely to be significant.

Keywords: Food price behavior; Oil price; Asymmetry; ARDL; Malaysia

JEL classification: C22; E31

Introduction

The episodes of rising food prices witnessed in recent years, especially during 2007–2008 and 2010–2011, have stimulated extensive popular and academic discussions and placed governments on alert as to their socio-economic implications[a]. With the need to carve or suggest policy prescriptions to contain food price escalation, numerous studies have been undertaken to identify determinants of food price variations. It is not surprising that, with concurrent upswings of crude oil price during the same years, the oil price has been examined as a potential explanation. A conventional wisdom tells us that, by affecting energy-intensive inputs such as fertilizers and fuel and influencing transportation costs, the oil price changes directly affect food production costs and subsequently food prices. In addition, given the increasing costs of global food production, food import bills would surge during times of rising oil price for food-importing countries and accordingly further exert an upward pressure on domestic prices of food items. Finally, the food versus fuel debate also links food price to oil price through the increase in food demand for the production of biofuels.

While the positive link between oil and food prices is well founded, existing empirical evidence is far from being uniform. Such studies as Zhang and Reed (2008), Zhang et al. (2010), Lambert and Miljkovic (2010), Nazlioglu and Soytas (2011), and more recently Reboredo (2012) have suggested the neutrality or, at best, only marginal reaction of various agricultural or food prices to oil price fluctuations. Evaluating the agricultural price (i.e. corn, soy meal and pork) uptrend in China from January 2000 to October 2007, Zhang and Reed (2008) side-line the oil price as a major underlying factor. By the same token, farm wages and manufacturing wages rather than fuel prices are noted to account for variations in the US food prices from 1970 to 2009 by Lambert and Miljkovic (2010). Further evidence is provided by Nazlioglu and Soytas (2011) for the case of Turkey. More specifically, examining the relations between oil price, lira-dollar exchange rate and individual agricultural prices (wheat, maize, cotton, soybeans and sunflower) using monthly data from January 1994 to March 2010, they provide evidence for the neutrality of these prices to oil price changes. This finding reaffirms earlier results for the global commodity prices by Zhang et al. (2010), who document the absence of long run relations and short run interactions between corn, rice, soybeans, sugar and wheat prices and oil price, and it is in line with the recent finding by Reboredo (2012), who notes that the food price spikes experienced in recent years are not caused by drastic increases in oil price. By contrast, Baffes (2007), Harri et al. (2009), Chen et al. (2010), and most recently Baffes and Dennis (2013) all provide evidence indicating significant contribution of oil price to agricultural prices[b].

These contrasting findings have continued to excite intense debate and paved ways for further research. Obviously, proper understanding the relations between domestic food prices and oil price for a country is most directly relevant for welfare assessment. However, for proper policy prescriptions to suppress any emerging food price crisis, any analysis of the oil price pass-through to domestic food prices must be cognizant at least indirectly of various underlying domestic factors such as market structure, public regulations, and cost structures. Recognition of these factors may hint on appropriate modelling strategies. More specifically, the presence of market power has normally been viewed to account for asymmetric price behavior with the adjustment to be quicker in the upward direction (Meyer and Cramon-Taubadel, 2004). Adding to this, we may also note the role of public policy schemes such as price floor and price ceiling in influencing asymmetric price behavior since they place the limit to which the price can adjust. Finally, the interplay between firms' cost structures and market power may account for both long run and short run asymmetries in the price movements, as explained by Karantininis et al. (2011a, b).

In this paper, we take part in this stream of research by examining the explanatory role of oil price in food price development from Malaysia's experience and perspectives. Malaysia is a fast-growing Asian economy underpinned by its structural transformation from a commodity-based economy to an industrial-based economy. The transformation has seen Malaysia to be increasingly reliant on food imports for its consumption. Being heavily dependent on food imports, Malaysia is arguably more exposed to oil and global food crises. In the year of escalating food and commodity prices in 2008, Malaysia's food import bills went up by more than 19%, far exceeding the annual average increase in the import bills of 8.5% from 2001 to 2010[c]. Accordingly, as regards to food security, Malaysia is likely to be at stake in the face of rising oil price. Further, in Malaysia, Yeong-Sheng (2008) computes the food budget across income quartiles using the Household Expenditure Survey 2004/2005 to be 33.03%, 25.92%, 21.2% and 14.63% for respectively Quartile 1, Quartile 2, Quartile 3, and Quartile 4. This means that the rising food prices will affect households at lower income quartiles more than those at the upper income levels. In light of recent energy market developments, proper assessment of the food price dynamics and its relations to oil price is urgently needed and hence the present analysis.

We adopt an alternative econometric framework, namely the nonlinear autoregressive distributed lags (NARDL) model recently advanced by Shin et al. (2011). We content that, in light of the forgoing discussion, the framework is most appropriate since it allows potential long-run and short-run asymmetries in the food price – oil price relations and hence indirectly hints on the importance of market power and policies in the food price dynamics. The rest of the paper is structured as follows. As a precursor to our analysis, the next section provides some background information and mentions some related studies on Malaysia. Then, Section 3 outlines the empirical approach. Data and results are discussed in Section 4. Finally, Section 5 concludes with the main findings and some concluding remarks.

Background

Malaysia's economic performance since independence in 1957 has been commendable, as manifested by its growth and inflation performance[d]. Over the span of more than 40 years from 1971 to 2012, Malaysia has experienced an annual average growth rate of 6.1%. The high growth performance of Malaysia, however, has also been marked by several setbacks, notably by the recessions of 1985 and 1998. After recording the average growth rate of 7.4% per year from 1971 to 1984, the real GDP contracted by slightly more than 1% in 1985. During 1987–2006, Malaysia witnessed miraculous growth performance registering the average growth rate of more than 8.5% per year. The one-decade long of high growth performance, however, was interrupted by the eruption of the Asian financial crisis in mid-1997. In 1998, Malaysia's real GDP contracted by more than 7% after recording the growth rate of more than 7% in 1997. Malaysia resumed its positive growth pattern again after the crisis. Despite recurring political and financial uncertainties since the Asian crisis with the latest global financial crisis being the notable example, Malaysia managed to record the average GDP growth rate of 5.0% during 1999–2012. In parallel, Malaysia has also recorded low inflation rate especially since 1985. Prior to 1985, i.e. 1971–1984, the average yearly inflation rate in Malaysia is 6.0%. Meanwhile, the corresponding figure for the period 1986–2012 is 2.6%.

These favourable indicators notwithstanding, the recent sharp swings in the oil price have generated a great deal of concern especially on the effects of oil price or the global food price on the domestic food price. To place this concern into perspectives, Figure 1 graphs the consumer price inflation as well as food price inflation from 1971–2012. As may be observed from the figure, despite low inflationary environment, Malaysia experienced several spikes of inflation, most notably in 1974 and 1981 where inflation peaked respectively at 16% and 9%. These two high inflation episodes are normally associated with the oil shocks of 1973 and 1979. Accordingly, with the recent sharp swings in the oil price, the inflationary effect of oil price has again taken a central stage in popular and policy discussion. Adding to this, the food price has exhibited higher volatility over the years with its upswing to be relatively steeper despite various policy mechanisms such as price controls and subsidies in place. This means that the welfare of especially lower groups of income is more likely to be at stake given their higher budget share on food items (Yeong-Sheng, 2008), an issue that requires immediate treatment.

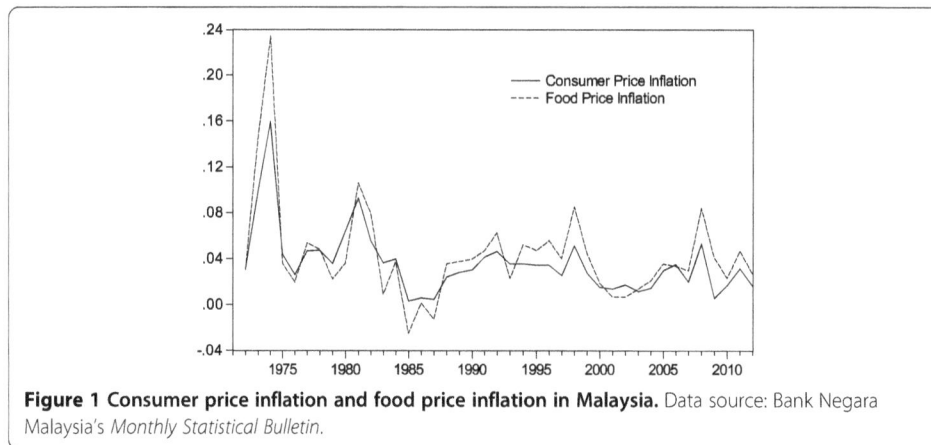

Figure 1 Consumer price inflation and food price inflation in Malaysia. Data source: Bank Negara Malaysia's *Monthly Statistical Bulletin.*

In addition, the growth performance of Malaysia has been underlined by its structural transformation from a commodity-based economy to an industrial-based economy. Over the years, Malaysia has become more and more dependent on food imports for the purpose of consumption. In Figure 2, we graph the nominal food imports in million ringgit (left axis) and oil price in ringgit (right axis). The nominal food imports have steadily increased over the years in parallel with the increase in real GDP. However, despite lower GDP growth in later years, the food import bills have been escalating especially since 2003, the period when the oil price witnessed a sharp increase. Hence, with the rising food import bills, the concern over food price inflation is well placed.

For Malaysia, empirical studies on inflation virtually centre on the aggregate price inflation and most of these studies predominantly look at the role played by money supply. See Ibrahim (2010), Tang (2010) and references therein. Recently, in the wake of the oil price hikes, several studies have emerged and evaluated the influences of oil price to domestic inflation. These include Cunado and de Gracia (2005), Jongwanich and Park (2009, 2011), and Ibrahim and Said (2012). Among these studies, only Ibrahim and Said (2012) consider the implications of oil price on various disaggregated consumer prices including the food price.

Cunado and de Gracia (2005) analyse the real and inflationary effects of oil price shocks for six Asian countries including Malaysia within the Granger causality framework. In the analysis, various non-linear transformations of oil price are employed.

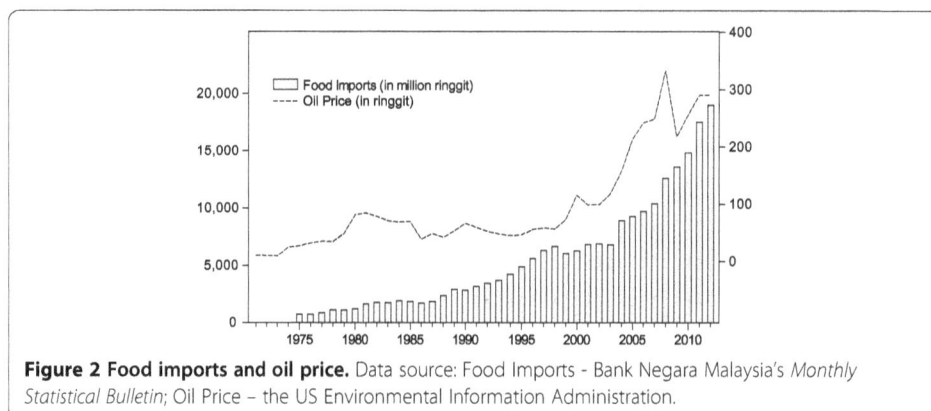

Figure 2 Food imports and oil price. Data source: Food Imports - Bank Negara Malaysia's *Monthly Statistical Bulletin*; Oil Price – the US Environmental Information Administration.

They find evidence that oil price shocks anticipate future variations in output growth and inflation especially when the oil price is expressed in local currencies. For the case of Malaysia, when the US dollar oil price is used, none of the oil price measures is found to Granger cause inflation. Using the oil price in ringgit, they also document the lack of Granger causality that runs from the oil price changes to inflation. However, the positive changes in oil price as well as the net oil price increase over 12 quarters tend to anticipate future variations in inflation. It should be noted that they do not compute the pass-through parameter and with the use of variables in first difference, the documented relations are short run in nature.

Contrasting these findings, Jongwanich and Park (2009) marginalize the role of oil and food price shocks in the inflation behavior of nine Asian countries including Malaysia. Aiming at evaluating the relative importance of various sources of inflation, they adopt a vector autoregressive (VAR) framework. According to them, the Asia's inflation is domestically driven. The external price shocks such as oil and food price shocks exert only minimal influences on the Asia's inflation performance. In their subsequent analysis, Jongwanich and Park (2011) estimate the pass-through coefficients of oil and food price shocks of these nine countries and conclude that the pass-through has been limited. In the case of Malaysia, the pass-through coefficient of oil price to producer prices is estimated to be roughly 0.14 while the pass-through to the consumer prices to be 0.025. They attribute the results to such government policy measures as subsidies and price controls.

More recently, Ibrahim and Said (2012) assess the oil price pass-through to various disaggregated consumer prices including the food price using a dynamic error-correction modelling. They also examine the asymmetric effects of positive and negative oil price changes in the analysis. For the consumer prices, the estimated long run oil price pass-through is 0.027, which is in close correspondence with that by Jongwanich and Park (2011). The long run pass-through to the food price is noted to be even higher, i.e. 0.056. In addition, they also provide evidence for the significant influences of only positive oil price changes in the short run. Both Cunado and de Gracia (2005) and Ibrahim and Said (2012) hint on the presence of asymmetry in the relations between consumer/food prices and oil price. However, their analyses are restricted to short run asymmetry only. We complement these studies by looking at both long run and short run asymmetries.

Methods

In the literature, the oil price – food price relations are normally examined by means of the standard time series techniques of cointegration, error-correction modelling and Granger causality. While the techniques enable evaluation of their long-run relations as well as their short-run interactions, they presume symmetric relations between agricultural prices and oil price. Accordingly, they are not adequate to capture potential asymmetries in the food price dynamics arising from among others the presence of market power and public policy schemes, as noted earlier. Recently, Shin et al. (2011) advance a nonlinear ARDL cointegration approach (NARDL) as an asymmetric extension to the well-known ARDL model of Pesaran and Shin (1999) and Pesaran et al. (2001), to capture both long run and short run asymmetries in a variable of interest. We adopt this modelling approach for our purpose[e].

To begin, we specify the following asymmetric long-run equation of food price (Schorderet, 2003 and Shin et al., 2011):

$$fp_t = \alpha_0 + \alpha_1 y_t + \alpha_2 op_t^+ + \alpha_3 op_t^- + e_t \tag{1}$$

where fp is food price, y is real income to capture aggregate demand or business cycle effect, op is oil price, and $\alpha = (\alpha_0, \alpha_1, \alpha_2, \alpha_3)$ is a cointegrating vector or a vector of long run parameters to be estimated. In (1), op_t^+ and op_t^- are partial sums of positive and negative changes in op_t:

$$op_t^+ = \sum_{i=1}^{t} \Delta op_i^+ = \sum_{i=1}^{t} \max(\Delta op_i, 0) \tag{2}$$

and

$$op_t^- = \sum_{i=1}^{t} \Delta op_i^- = \sum_{i=1}^{t} \min(\Delta op_i, 0) \tag{3}$$

Based on the above formulation, the long run relation between food price and oil price increases is α_2, which is expected to be positive. Meanwhile, α_3 captures the long run relation between food price and oil price reduction. Since they are expected to move in the same direction, α_3 is expected to be positive. We further posit that the oil price increases will result in higher long run changes in the food price as compared to the food price impact of oil price reduction of the same magnitude, i.e. $\alpha_2 > \alpha_3$. Thus, the long run relation as represented by (1) reflects asymmetric long-run oil price pass-through to the food price.

As shown in Shin et al. (2011), equation (1) can be framed in an ARDL setting along the line of Pesaran and Shin (1999) and Pesaran et al. (2001) as:

$$\begin{aligned}
\Delta fp_t = &\alpha + \beta_0 fp_{t-1} + \beta_1 y_{t-1} + \beta_2 op_{t-1}^+ + \beta_3 op_{t-1}^- + \sum_{i=1}^{p} \phi_i \Delta fp_{t-i} \\
&+ \sum_{i=0}^{q} \gamma_i \Delta y_{t-i} + \sum_{i=0}^{s} (\theta_i^+ \Delta op_{t-i}^+ + \theta_i^- \Delta op_{t-i}^-) + u_t
\end{aligned} \tag{4}$$

Where all variables are as defined above, p, q and s are lag orders and $\alpha_2 = -\beta_2/\beta_0$, $\alpha_3 = -\beta_3/\beta_0$, the aforementioned long run impacts of respectively oil price increase and oil price reduction on the food price. $\sum_{i=0}^{s} \theta_i^+$ measures the short-run influences of oil price increases on food price inflation while $\sum_{i=0}^{s} \theta_i^-$ the short run influences of oil price reduction on food price inflation. Hence, in this setting, in addition to the asymmetric long run relation, the asymmetric short-run influences of oil price changes on food price inflation are also captured.

Empirical implementation of the nonlinear ARDL approach entails the following steps. First, while the ARDL approach to cointegration is applicable irrespective of whether the variables are I(0) or I(1), it is still necessary to conduct unit root tests such that no I(2) variable is involved. This is important since the presence of an I(2) variable renders the computed F-statistics for testing cointegration invalid. To this end, we apply the widely-used ADF and PP unit root tests for establishing the variables' orders of integration. In the second step, we estimate equation (4) using the standard OLS estimation method. As in Katrakilidis and Trachanas (2012), we adopt the general-to-specific procedure to arrive at the final specification of the NARDL model by trimming insignificant lags. Third, based on the estimated NARDL, we perform a test for the presence of cointegration among the variables using a bounds testing approach of Pesaran et al. (2001) and Shin et al. (2011). This involves the Wald F test of the null hypothesis, $\beta_0 = \beta_1 = \beta_2 = \beta_3 = 0$. In the final step, with the presence of cointegration, examination of long-run and short-run asymmetries in the relations between oil and food prices is made and inferences are drawn. In this step, we can also derive the asymmetric cumulative dynamic multiplier effects of a one percent change in op_{t-1}^{+} and op_{t-1}^{-} respectively as:

$$m_h^{+} = \sum_{j=0}^{h} \frac{\partial y_{t+j}}{\partial op_{t-1}^{+}}, \quad m_h^{-} = \sum_{j=0}^{h} \frac{\partial y_{t+j}}{\partial op_{t-1}^{-}}, \quad h = 0, 1, 2 \dots \quad (5)$$

Note that as $h \to \infty$, $m_h^{+} \to \alpha_2$ and $m_h^{-} \to \alpha_3$.

Results and Discussion

We employ annual data from 1971 to 2012 in the analysis. The food price index corresponding to the food price component of the consumer price index is used to capture the food price (*fp*) in Malaysia. The real income is represented by real gross domestic product (*y*). For the oil price (*op*), the West Texas intermediate crude oil price in ringgit, i.e. $op_{rm} = op \times rm$, is used, where *rm* is the ringgit exchange rate vis-à-vis the US dollar. In their analysis of several Asian countries, Cunado and de Gracia (2005) find that the inflationary effect of oil price hikes is more prevalent when the oil price is expressed in domestic currencies. Still, it may be argued that, while the changes in the exchange rate is one of potential channels of the oil price pass-through to domestic prices, the changes in the exchange rate can also be driven by some other factors not related to oil prices. Accordingly, multiplying the oil price with the exchange rate to arrive at the ringgit-denominated oil price may erroneously attribute the pass-through to the oil price when the exchange rate changes are due to non-oil factors. To see this, we plot in Figure 3 the evolution of the ringgit-USD exchange rate together with the oil price in the US dollar. In early years and later years of the sample, oil price increases seem to be accompanied by the ringgit appreciation (or the USD depreciation against

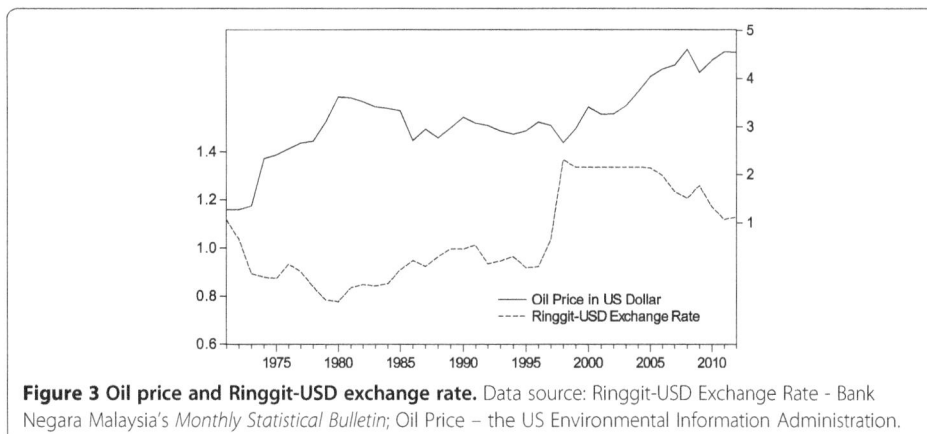

Figure 3 Oil price and Ringgit-USD exchange rate. Data source: Ringgit-USD Exchange Rate - Bank Negara Malaysia's *Monthly Statistical Bulletin*; Oil Price – the US Environmental Information Administration.

the ringgit). However, when the ringgit depreciated during 1980s, the oil price was relatively stable. The oil price was also relatively stable during the drastic depreciation of the ringgit in 1997–1998 when Malaysia suffered from the Asian financial crisis. As we are aware, the drastic downfall of the ringgit value stemmed from the crisis. Hence, converting the oil price into the ringgit may erroneously pick up the effect of exchange rate depreciation on food price inflation. As an additional exercise, we also investigate the sensitivity of the results to the employment of the oil price in the US dollar $(op_{usd})^f$. All variables are expressed in natural logarithm. Except the oil price, the data are sourced from the *Monthly Statistical Bulletin* of Malaysia's Central Bank (www.bnm. gov.my). The oil price is taken from the US Environmental Information Administration (www.eia.gov).

Given the requirement of the bounds testing procedure that no I(2) variables are involved, we first subject each time series to the ADF and PP unit root tests. The results of these tests are given in Table 1. In the tests, we include both constant and trend terms and employ the SIC for the optimal lag order in the ADF test equation. Both ADF and PP unit tests are in agreement that real GDP and the two oil price measures are integrated of order 1. However, for the food price, the ADF test indicates its stationarity in level while the PP test suggests that it becomes stationary after first differencing. Since the tests indicate none of the variables is I(2), we can proceed to the bounds testing procedure.

Accordingly, we estimate equation (4) and apply the general-to-specific procedure to arrive at the model final specification. The maximum lag order considered is 3. Table 2 reports the bounds F-statistics and Table 3 present the model estimation results. From the bounds F-statistics, we come to the conclusion that the three variables, i.e. food price, real income and oil price, co-move in the long run. The statistics, 11.90 and 12.48 for respectively equations with oil price in ringgit and oil price in the US dollar, exceed the critical upper bound. With this finding, we are in position to assess the food price dynamics and its relation to real GDP and positive and negative changes in oil price.

Before inferences are drawn, we first judge the adequacy of the dynamic specification on the basis of various diagnostic statistics. These include the Jarque-Bera statistics for error normality (J-B), the LM statistics for autocorrelation up to order 2, and the ARCH statistics for autoregressive conditional heteroskedasticity up to order 2. These are presented at the lower panel of Table 3. In addition, we also graph the CUSUM and CUSUMSQ statistics for testing structural stability of the model in Figure 4. The model with the oil price in ringgit passes all diagnostic tests suggesting error normality,

Table 1 ADF and PP unit root tests

Variable	Level		First difference	
	ADF	PP	ADF	PP
fp	−4.6516***	−3.5109	−4.5033***	−4.5938***
y	−1.5225	−1.6961	−5.5803***	−5.5819***
op_{rm}	−2.1053	−2.1156	−6.4388***	−6.4406***
op_{usd}	−2.2269	−2.2275	−6.2351***	−6.2351***

Notes: the constant and trend terms are included in the test equations and the SIC is used to select the optimal lag order in the ADF test equation.
***denote significance at 1% level.

Table 2 Bounds test for nonlinear cointegration

Oil price specification	F-Statistics	95% lower bound	95% upper bound	Conclusion
op_{rm}	11.9022	4.428	6.250	Cointegration
op_{usd}	12.4833			Cointegration

Notes: the critical values are from Narayan (2005), given the small sample size.

absence of autocorrelation and ARCH effect, and parameter stability. Likewise, the model with the oil price in the US dollar also passes all these tests except the ARCH test. Accordingly, the dynamics of food price inflation is adequately specified.

From the estimated results in Table 3, we compute the cointegrating and long-run equations for both models. These are presented in Table 4. The long run coefficients of real income are positive and significant at 1% significance level, as should be expected. They suggest that a 1% increase in real income is related to the increase in the expected food price inflation by roughly 0.50%, holding the oil price constant. Turning to our main theme, we note the asymmetric long run relation between the food price and oil price with the increase in oil price being significantly related to the food price while the reduction in oil price not. In line with many studies, the pass-through of oil price to the food price is not complete. Our estimates suggest that a 10% increase in the price of oil is associated with the increase in the expected food price in the range of 0.6% to 0.8%. From the estimates, the use of oil price in ringgit or the oil price in US dollar results in no marked difference in the long run oil price pass through. We believe that a slightly higher magnitude of the oil price pass-through when the oil price is expressed in the ringgit may pick up the effect of ringgit depreciation especially during the episode of the 1997/1998 Asian financial crisis.

Table 3 Nonlinear ARDL estimation results

Independent variable	Oil price specification				
	Oil price in Ringgit			Oil price in USD	
	Coefficient	p-vale		Coefficient	p-value
Constant	−1.0406	0.0239		−1.1539	0.0100
$p(-1)$	−0.5592	0.0000		−0.5082	0.0000
$y(-1)$	0.2690	0.0000		0.2628	0.0000
$op^+(-1)$	0.0458	0.0041		0.0307	0.0355
$op^-(-1)$	0.0293	0.2136		0.0216	0.2971
$\Delta p(-1)$	0.3549	0.0397		0.2385	0.0216
Δop^+	---	---		0.0694	0.0044
$\Delta op^+(-1)$	−0.0514	0.0397		---	---
R^2	0.6798			0.7098	
J-B	1.4480	0.4848		1.1050	0.5755
LM(1)	0.0022	0.9622		0.1303	0.7181
LM(2)	0.3534	0.8380		0.4439	0.8010
ARCH(1)	0.0040	0.9496		4.1510	0.0416
ARCH(2)	0.5286	0.7677		9.4405	0.0089

Notes: J-B is the Jarque-Bera test for error normally, LM(.) is the LM test for error autocorrelation up to the lag order given in the parenthesis, and ARCH(.) is the ARCH test for autoregressive conditional heteroskedasticity up to the lag order given in the parenthesis.

(a) Model with oil price in domestic currency

(b) Model with oil price in US dollar

Figure 4 CUSUM and CUSUMSQ.

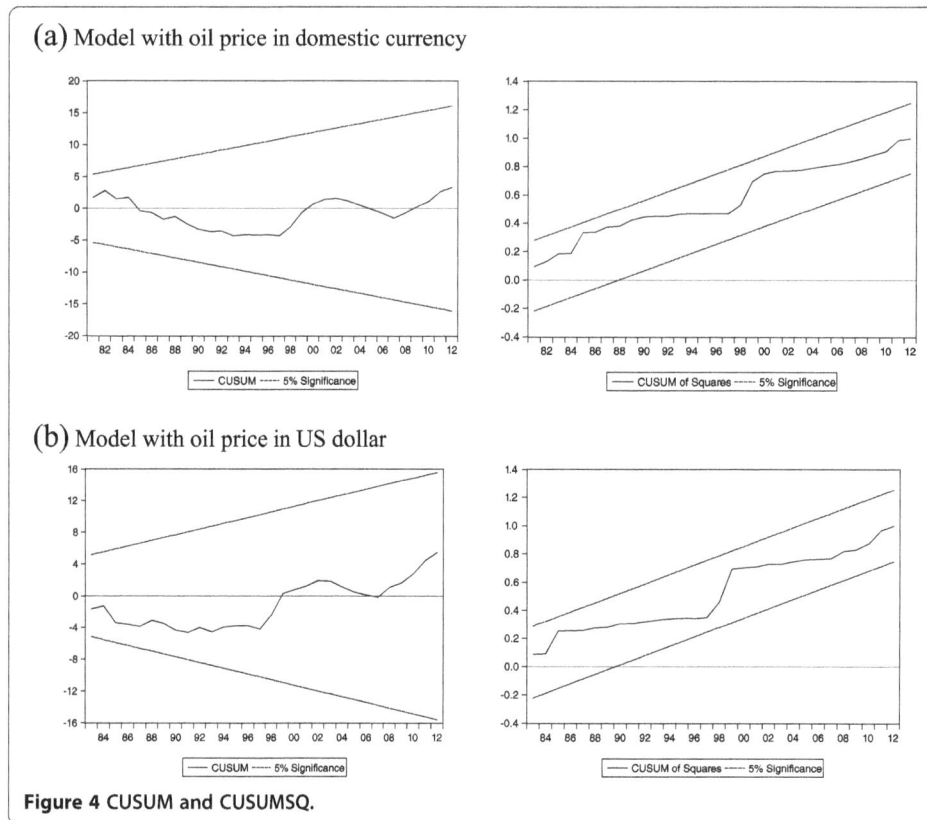

While the estimates we obtain can be viewed as low, we believe that they are sensible. In the analysis of the global food price, the long run pass-through coefficient is estimated to be 0.18 by Baffes (2007). It should be expected that the effect of oil price should be lower as we go down the supply chain. Hence, it is reasonable to observe lower pass-through to the domestic food price. Moreover, in Malaysia, the prices of several essential food items are controlled and administered by the government. This may have shaped the low oil price – food price relation in the long run, in line with the argument forwarded by Jongwanich and Park (2011).

This low pass-through of oil price increase notwithstanding, it is a cause of concern especially to households at lower income quartiles. Over the past years, the upswings in the oil price have been sharp and, hence, can contribute quite substantially to the food price increase. As an example, in the year of drastic increase in the oil price in ringgit in 2008, the oil price jumped by almost 30% from the preceding year. On the basis of our estimates, the oil price would account for the increase of roughly 2.4% in the food

Table 4 Long-run relations

Variable	Oil price specification			
	Oil price in Ringgit		Oil price in USD	
	Coefficient	p-vale	Coefficient	p-value
Constant	−1.8607	0.0135	−2.2707	0.0075
y	0.4810	0.0000	0.5171	0.0000
op^+	0.0819	0.0021	0.0605	0.0141
op^-	0.0524	0.2053	0.0425	0.2918

price over the long run. This is high considering the average annual food price inflation in Malaysia of 4.3%. Thus, given higher budget shares of households at quartile 1 (33.03%) and quartile 2 (25.92%), these groups would be more affected by the oil price increases than those at the upper income levels would. Moreover, as can be further noted from the Table, the long run relation between food price and oil price reduction is insignificant. This finding should be worrying since, while the drastic oil price increase is positively related to food price in the long run, its decline will not be translated into a reduction in the food price. In other words, the high food price will linger around even if the oil price has corrected downward after the initial increase.

As for the short run, our results provide evidence for the presence of asymmetry as well. From Table 3, we may observe only the significance of the positive change in the oil price increase (i.e. Δop^+). However, the sign of Δop^+ tends to be perverse depending on the model used. In the model with the US dollar-denominated oil price, the change in the oil price increase is contemporaneously and significantly related to the food price inflation. Its coefficient is positive suggesting an impact/immediate increase in the food price inflation by 0.07 percentage point. However, when the oil price in ringgit is used, the once-lagged Δop^+ turns out to be significant and it is negatively signed. One potential explanation is, in the latter, the interplays between the oil price and the ringgit exchange rate may have accounted for the result. We note that, after the de-pegging of the ringgit in July 2005, the ringgit has exhibited an appreciation trend. The period after 2005 has also been marked by the uptrend and sharp swings in the oil price. Apart from these results, we also find some persistence in the food price inflation as reflected by the positive and significant coefficient of once-lagged inflation rate. However, in the short run, real income does not seem to exert any causal influences on the domestic food inflation.

We also compute the short-run and long-run multipliers of oil price increases on the food price. These are given in Figure 5[8]. Corresponding multipliers of oil price reduction are not measured given the insignificant effects of oil price reduction both in the short run and in the long run. Note that, it takes roughly 6 to 7 years for the impact of initial oil price increases to be fully felt; that is, it converges to the long run estimate of 0.0819 after 6 years.

While the sources of asymmetries are not addressed directly in the present paper, the documented patterns of asymmetries tend to hint on the presence of market power

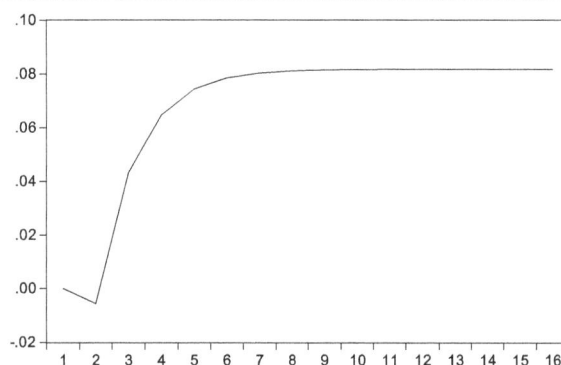

Figure 5 Short-run and long-run multipliers of oil price increases.

within the food supply chains. As noted by Meyer and Cramon-Taubadel (2004) and discussed in Karantininis et al. (2011a, b), the scale and adjustment costs are the likely sources of short-run asymmetry while the market power the source of long-run asymmetry. Thus, the evidence that the food price is directly related to the increase in the oil price and not to its reduction in the long run provides indication that suppliers may have exercised their market power. Then, the complete absence of the short-run influences of negative oil price changes on the food price further strengthens the conclusion that the market power does play a significant role (Karantininis et al. 2011a, b). On this basis, the recent passage of profiteering and anti-competition acts is a step in the right directions. However, its implementation should not be limited to retailers but should be applied broadly to all suppliers in the supply chain. Moreover, the benefits of such public policy schemes as price controls of essential food items, subsidies and stockpiling are likely to be limited and to work only in the short run unless the issue of market power is properly tackled.

Conclusion

In this study, we examine the bearings oil price has on the food price index for a net food-importing country, i.e. Malaysia. Recognizing potential roles of market power, cost structures and public policy schemes in shaping asymmetric behavior of the food price, we adopt a nonlinear ARDL model for the analysis to capture both long-run and short-run asymmetric relations between the food and oil prices. From the analysis, we find evidence for the presence of asymmetries in the long run as well as in the short run. More specifically, in the long run, an oil price increase tends to lead to the increase in the food price while the oil price reduction does not seems to be related to food price. From the estimated cointegrating vectors, we note the degree of pass-through from the oil price increase to the food price to be low and far from complete. Likewise, in the short run, only the changes in the oil price increase are significantly related to the food price inflation. These results are largely similar regardless of whether the oil price in ringgit or the oil price in the US dollar is used.

The low oil price pass-through may be attributed to such public policy schemes as administered prices of essential food items and subsidies as well as adjustment costs. Still, it is a cause of concern especially to those at lower income levels. Over the years, various approaches have been adopted by the government of Malaysia to address rising oil price. These include enhancement of agricultural productivity, betterment of land management, improvement of distribution and marketing chains, effective price administration, stockpiling and subsidies. The recent anti-profiteering and anti-competition acts as regulatory mechanisms to curtail rising domestic prices are also notable.

Our results hint on several aspects of these approaches that need further attention. First, the evidence that only oil price increase and changes in the oil price increase are significantly related to the food price is likely to indicate the presence of market power in the Malaysian food markets. Thus, policy attention should be directed to containing market power and, to be more effective, should cover all suppliers (importers, wholesalers and retailers) in the food supply chain. We believe that the recent profiteering and anti-competition acts are a regulatory step in the right direction. As such, enforcement of these acts must be strengthened. Second, although the persistent budget deficit experienced since 1998 has results in rationalization of oil subsidies on several occasions, it must be adopted gradually and together with mechanisms to enhance competition in the domestic food markets and to increase domestically produced food such that there will be minimal impacts on the poor. Finally, the benefits of such existing policy schemes as price controls of essential food items, subsidies and stockpiling are likely to be limited and to work only in the short run unless the issue of market power and agricultural productivity are properly addressed.

Endnotes

[a]Among the much debated implications of food price increases are malnutrition and poverty (Ivanic et al., 2012, Zheng and Henneberry, 2012, Anriquez et al., 2013, Fujii, 2013, Wood et al. 2012).

[b]The list of studies on oil and commodity price links is extensive and we mention only few representative studies here. Interested readers may refer to Serra and Zilberman (2013) for a survey.

[c]Food imports refer to nominal imports of primary and processed food and beverages taken from the Monthly Statistical Bulletin published online by Bank Negara Malaysia (www.bnm.gov.my).

[d]All figures in this section are calculated based on data from the *Monthly Statistical Bulletin* of Malaysia's Central Bank, i.e. Bank Negara Malaysia.

[e]Among studies that have adopted the NARDL method include Katrakilidis and Trachanas (2012) for house price dynamics in Greece, Delatte and Lopez-Villavicencio (2012) for the exchange rate pass-through in four major developed economies, and Verheyen (2013) for the non-linear influence of exchange rates on EMU exports to the US. A notable study related to food price is Karantininis et al. (2011a, b) for the Swedish pork market.

[f]We also include the exchange rate in the latter model. However, the exchange rate turns out to be insignificant both in the long run and in the short run while other results remain largely similar. Accordingly, we do not report these results to conserve space.

[g]The calculation is based on the model with oil price in Ringgit as reported in Table 3.

Competing interests
The author declares that he has no competing interests.

Acknowledgements
I would like to thank anonymous referees of the journal for providing constructive comments on the paper. However, any remaining error is my own responsibility.

References
Anriquez G, Daidone S, Mane E (2013) Rising food prices and undernourishment: a cross-country inquiry. Food Policy 38:190–202

Baffes J (2007) Oil spills on other commodities. Resource Policy 32(3):126–34

Baffes J, Dennis A (2013) Long-Term Drivers for Food Prices, The World Bank Policy Research Work Paper No. 6455

Chen S-T, Kua HI, Chen C-C (2010) Modelling the relationship between the oil price and global food prices. Appl Energy 87(7):2517–25

Cunado J, de Gracia FP (2005) Oil prices, economic activity and inflation: evidence from some Asian countries. Q Rev Econ Finance 45(1):65–83

Delatte A-L, Lopez-Villavicencio A (2012) Asymmetric exchange rate pass-through: evidence from major countries. J Macroeconomics 34:833–44

Fujii T (2013) Impact of food inflation on poverty in the Philippines. Food Policy 39:13–27

Harri A, Nally L, Hudson D (2009) The relationship between oil, exchange rates, and commodity prices. J Agric Appl Econ 41(2):501–10

Ibrahim MH (2010) Money-price relation in Malaysia: has it disappeared or strengthened? Econ Changes Restruct 43(4):303–22

Ibrahim MH, Said R (2012) Disaggregated consumer prices and oil price pass-through: evidence from Malaysia. China Agric Econ Rev 4(4):514–29

Ivanic M, Martin W, Zaman H (2012) Estimating the short-run poverty impacts of the 2010–11 surge in food prices. World Dev 40(11):2302–17

Jongwanich J, Park D (2009) Inflation in developing Asia. J Asian Econoomics 20(5):507–18

Jongwanich J, Park D (2011) Inflation in developing Asia: pass-through from global food and oil price shocks. Asian-Pacific Econ Lit 25(1):79–92

Karantininis K, Katrakylidis K, Persson M (2011a) Price Transmission in the Swedish Pork Chain: Asymmetric Non Linear Ardl. EAAE 2011 Congress, Zurich, Switzerland

Karantininis K, Kostas K, Persson M (2011b) Price transmission in the Swedish pork chain: Asymmetric nonlinear ARDL., Paper presented at the EAAE 2011 Congress: Challenges and Uncertainty, September 2, 2011, Zurich, Switzerland

Katrakilidis C, Trachanas E (2012) What drives housing price dynamics in Greech: new evidence from asymmetric ARDL cointegration. Econ Model 29:1064–9

Lambert DK, Miljkovic D (2010) The sources of variability in US food prices. J Policy Model 32:210–22

Meyer J, Cramon-Taubadel S (2004) Asymmetric price transmission: a survey. J Agric Econ 55(3):581–611

Narayan PK (2005) The saving and investment nexus for China: evidence from cointegration tests. Appl Econ 37(17):1979–90

Nazlioglu S, Soytas U (2011) World oil prices and agricultural commodity prices: evidence from an emerging market. Energy Economics 33(3):488–96

Pesaran MH, Shin Y (1999) An autoregressive distributed lag modelling approach to cointegration analysis. In: Storm S (ed) Econometrics and Economic Theory in the 20th Century: The Ragnar Frisch Centennial Symposium, Chapter 11. Cambridge University Press, Cambridge

Pesaran MH, Shin Y, Smith RJ (2001) Bounds testing approaches to the analysis of level relationship. J Appl Econometrics 16:289–326

Reboredo JC (2012) Do food and oil prices co-move? Energy Policy 49:456–67

Schorderet Y (2003) Asymmetric Cointegration. Working Paper. Department of Economics, University of Geneva

Serra T, Zilberman D (2013) Biofuel-related price transmission literature: a review. Energy Economics 37:141–51

Shin Y, Yu B, Greenwood-Nimmo M (2011) Modelling Asymmetric Cointegration and Dynamic Multiplier in a Nonlinear ARDL Framework, Mimeo

Tang CF (2010) The money-price nexus for Malaysia: new empirical evidence from the time-varying cointegration and causality tests. Glob Econ Rev 39(4):383–403

Verheyen F (2013) Exchange rate nonlinearities in EMU exports to the US. Econ Model 32:66–76

Wood BDK, Nelson CH, Nogueira L (2012) Poverty effects of food price escalation: the importance of substitution effects in Mexican households. Food Policy 37(1):77–85

Yeong-Sheng T (2008) Household Expenditure on Food at Home in Malaysia, MPRA Paper No. 15031

Zhang Q, Reed M (2008) Examining the Impact of the World Crude Oil Price on China's Agricultural Commodity Prices: the Case of Corn, Soybean, and Pork. The Southern Agricultural Economics Association Annual Meeting, Dallas, TX

Zhang Z, Lohr L, Escalante C, Wetzstein M (2010) Food versus fuel: what do prices tell us? Energy Policy 38(1):445–51

Zheng Z, Henneberry SR (2012) Estimating the impacts of rising food prices on nutrient intake in urban China. China Econ Rev 23(4):1090–103

7

Consumer demand for alcoholic beverages in Switzerland: a two-stage quadratic almost ideal demand system for low, moderate, and heavy drinking households

Matteo Aepli

Correspondence: aepli@ethz.ch
Agricultural Economics Group,
Institute for Environmental
Decisions, ETH Zurich,
Sonneggstrasse 33, SOL E5 8092
Zurich, Switzerland

Abstract

In this study, we estimate final demand for beverages with a particular focus on alcoholic beverages and calculate elasticities using microdata from the Swiss household expenditure survey from 2000 to 2009, which containsdata from more than 34,000 households. We estimate price and income responses for three household segments – light, moderate, and heavy drinking households – to assess whether higher alcohol consumption could be described by different price and income elasticities in comparison to lower alcohol consumption. We obtain unconditional estimates by applying a two-stage budgeting quadratic almost ideal demand system. To generate missing price data, we used the recently proposed quality adjusted price approach. Due to a high share of zero consumption for some beveragescategories,we correct the model with a two-step estimation procedure. Estimation results show that heavy drinking households are much less price elastic with respect to wine and beer in comparison to moderate or light drinking households, while the price response for spirits is almost constant over the three segments. Before implementing a new tax for alcoholic beverages in Switzerland, the social, health, and economic effects of a rather small decrease in alcohol consumption among heavy drinking households must be weighed against possible negativeconsequences of a sharp decline in light or moderate drinking households.

Keywords: Demand alcoholic beverages; QUAIDS; Household segments

Introduction

Knowledge about the determinants of consumption of alcoholic beverages and the price and income elasticity of different consumer segments is highly relevant to policy decisions. Alcohol consumption is a public health priority. Following for example Wakabayashi (2013) and Corrao et al. (2000), a regular but moderate consumption of alcohol in general can have a positive effect on health by increasing the level of HDL cholesterol and reducing the risk of heart and vascular diseases, though other studies do not report positive health effects of low or moderate alcohol consumption (Estruch et al. 2014, WHO 2007). The discussion is complex and still ongoing. However, there is consensus that excessive alcohol consumption can be detrimental to health in that it can lead to liver cancer, liver disease, higher blood pressure, stroke, or mental decline

(Rehn et al. 2001; Schwartz et al. 2013; Sabia et al. 2014). In addition, excessive alcohol consumption can also have negative psychological and behavioral consequences. It can lead to higher rates of crime or violence (Jacobs and Steyn 2013; Fergusson et al. 2013) and have negative economic effects, leading to substantial higher state costs (Sacks et al. 2013).

Policy measures have been shown to reduce alcohol consumption and to be a cost effective health care intervention (Xuan et al. 2013; Doran et al. 2013). Nevertheless, alcohol taxation and its effectiveness is still a controversial discussion point in Switzerland and other European countries (for an overview of the different alcohol policy regulations in Europe see e.g., FOPH, 2014a). While total alcohol consumption in Switzerland has been decreasing slightly since 1990, with a current per capita consumption of about 8.5 liters of pure alcohol per year (SAB 2013), excessive forms of alcohol drinking like binge drinking have been increasing, especially among young people (FOPH 2013; Annaheim and Gmel 2004). Therefore, Switzerland is currently debating a tightening of alcohol legislation. While wine is excluded from the alcohol tax, beer and spirits are heavily taxed. The current legislation has the purpose of a positive health effect on one hand and a fiscal benefit on the other, with annual tax revenue of about 440 million Swiss francs (FDF 2009).

In order to assess the effect of a new tax on consumption, especially in the case of heavy drinkers, the estimation of price elasticities is crucial. However, price policies may be superfluous if changing income level affects consumption (Aepli and Finger 2013; Gallet 2010). Therefore, an estimation of income elasticities as a supplement to price elasticities is necessary. An additional tax burden to reduce light or moderate drinkers' consumption levels may not be desirable with respect not only to social aspects but also concerning health effects, depending on which assumptions are made with respect to the health effects of low or moderate alcohol consumption. In the case of heavy drinkers it is clear that a reduction in consumption would reduce public health costs and decrease social problems. To address the question of how price and income affect demand for alcoholic beverages in Switzerland, we estimate final demand price and income elasticities separately for light, moderate, and heavy drinking households based on a repeated cross-sectional household expenditure survey. We combine several recently used methods into one demand system, expecting that price and income responses are not constant among the three segments. Switzerland is a particularly interesting case due to its comparatively high purchasing power. Findings from the Swiss case can also be transferred to a considerable extent to other European countries.

Elasticity estimates based on household data are scarce. Most studies rely on time-series data (Gallet 2007), which limits the ability to estimate elasticities for different household segments. Furthermore, most studies did not investigate possible substitution effects between alcoholic and non-alcoholic beverages. Our study contributes to filling this gap and allows a more detailed understanding of alcohol demand with respect to the demand response of different household segments.

We apply a two-stage quadratic almost ideal demand system (QUAIDS) (Banks et al. 1997 [for application see e.g., Abdulai 2002; Jithitikulchai 2011]). It uses cross-sectional data from the Swiss household expenditure survey from 2000 to 2009, which contains data from more than 34,000 households. Final demand elasticities are estimated for food, beverages, and other products and services at the first stage and for wine, beer, spirits, and several non-alcoholic beverages at the second stage. Due to a high zero consumption for some product groups, we modified the QUAIDS using Shonkwiler and Yen's approach (1999) (for application see e.g., Thiele 2008; Tafere et al. 2010) and corrected for possible heteroscedasticity by applying a parametric bootstrap. We received missing price data by using the method recently proposed by Majumder et al. (2012), further developed by Aepli and Finger (2013), and by introducing a variable for seasonality and trend applied by Aepli and Kuhlgatz (2014) for the same expenditure survey. Furthermore, we corrected for endogeneity of the expenditure variable in the model by using the augmented regression technique.

Heeb et al. (2003) analyzed the effect in Switzerland of a price reduction on spirits in 1999. The reduction was due to a tariff reduction under the WTO agreement and they focused on heavy drinkers using a longitudinal survey. Overall, spirits are rated as inelastic, whereas low volume drinkers are more elastic than high volume drinkers are. This would mean that taxes should reduce light or moderate drinkers' level of consumption more than the heavy drinkers' levels (Manning et al. 1995). Kuo et al. (2003) used the same period as Heeb et al. (2003) with the price reduction due to the WTO agreement to estimate demand reaction to the price decrease. They found that young people in particular responded with higher demand, whereas people aged 60 or older did not respond at all to the price change.

The paper is structured as follows. In Section Theoretical framework and two-stage budgeting, we describe the theoretical framework and the two-stage budgeting system, and in Section Data description and price computation, we present the descriptive statistics and indicate how price data are generated. Section Estimation procedure and elasticity calculation summarizes the estimation procedure and the elasticity calculation. The income and price responses are discussed in Section Results and discussion, while Section Conclusions and policy implications concludes.

Background

Theoretical framework and two-stage budgeting

Theoretical framework of the QUAIDS

Equation systems have been widely used in demand analysis. The most frequently applied models are: the almost ideal demand system (AIDS) or the linearized AIDS (Deaton and Muellbauer 1980a, 1980b; Akbay et al. 2007; Mhurchu et al. 2013); the quadratic version of the AIDS (QUAIDS) (Banks et al. 1997; Abdulai 2002; Dey et al. 2011; see Stasi et al. 2010 and Cembalo et al. 2014 for wine in particular); the Rotterdam model (Barten 1964; Theil 1965; Barnett and Seck 2008; Barnett and Kanyama 2013); the Translog (Christensen et al. 1975; Holt and Goodwin 2009); and the linear and quadratic expenditure systems (Pollak and Wales 1978; De Boer and Paap 2009). One of the most important criteria is the approximation of non-linear Engel curves, which is best satisfied by the QUAIDS and allows general income responses that are not captured by the AIDS or many other models. The QUAIDS model is a rank three demand system and satisfies the axioms of choice. It allows exact aggregation over consumers due to underlying preferences that are of the generalized Gorman polar form (Banks et al. 1997; Blackorby et al. 1978). The recently proposed EASI demand system (Lewbel and Pendakur 2009; for application see e.g., Stasi et al.

2011) would have been an alternative due to its flexibility with respect to the approximation of Engel curves. But Cranfield et al. (2003) noted that the QUAIDS is especially suitable in the case of a disaggregated analysis, as in this study. Furthermore, the QUAIDS has already been successfully applied to the Swiss household expenditure survey (see e.g., Abdulai 2002).

Following Banks et al. (1997) the indirect utility function of the QUAIDS is given by:

$$\ln V = \left(\left[\frac{\log m - \log a(p)}{b(p)} \right]^{-1} + \lambda(p) \right)^{-1} \tag{1}$$

where m is total expenditure and p are prices, $a(p)$ is the translog price aggregator, and $b(p)$ is the Cobb-Douglas price aggregator. The term $\frac{\log m - \log a(p)}{b(p)}$ is the indirect utility function of a PIGLOG demand system and $\lambda(p)$ is a differentiable, homogeneous function of degree zero of prices. $a(p)$ and $b(p)$ are defined as follows:

$$a(p) = \alpha_0 + \sum_{i=1}^{n} \alpha_i \log p_i + \frac{1}{2} \sum_{i=1}^{n} \sum_{j=1}^{n} \gamma_{ij} \log p_i \log p_j \tag{2}$$

$$b(p) = \beta_0 \prod_{i=1}^{n} p_i^{\beta_i} \tag{3}$$

where i and j are specific goods and n is the number of goods.

The Marshallian demand function in budget shares is obtained by applying Roy's identity to the indirect utility function:

$$w_i = \alpha_i + \sum_{j=1}^{n} \gamma_{ij} \log p_j + \beta_i \log \left[\frac{m}{a(p)} \right] + \frac{\lambda_i}{b(p)} \left\{ \log \left[\frac{m}{a(p)} \right] \right\}^2 + \varepsilon_i \tag{4}$$

where w_i is the budget share for product category i, and α_0, α_i, γ_{ij}, β_i and λ_i are parameters to be estimated. The residuals ε_i are assumed to be multivariate normal distributed with zero mean and a finite variance-covariance matrix.

To meet utility maximization theory, the restriction of adding-up (5), of homogeneity of the Marshallian cost function in prices and total expenditure (6), and of symmetry of the Slutsky matrix (Young's theorem) (7) are implemented:

$$\text{Adding-up: } \sum_{i}^{n} \alpha_i = 1; \sum_{j}^{n} \gamma_{ij} = 0; \sum_{i}^{n} \beta_i = 0; \sum_{i}^{n} \lambda_i = 0 \tag{5}$$

$$\text{Homogeneity: } \sum_{i=1}^{n} \gamma_{ij} = 0 \tag{6}$$

$$\text{Symmetry: } \gamma_{ij} = \gamma_{ji} \tag{7}$$

Theoretical restrictions allow us to reduce the number of estimated parameters as well as improve the efficiency of the estimated model (Barnett and Seck 2008). To avoid singularity in the variance-covariance matrix and to satisfy the condition of adding-up, one equation is dropped and an n-1 equation system is estimated. The parameters of the n th equation are obtained from the restriction and the parameters of the n-1 equations.

In addition to the economic determinants, we take household characteristics and a variable for seasonality and time trend into account using the demographic translation approach by Pollak and Wales (1978) (applied e.g., by Abdulai 2002; Bopape and Myers 2007), modifying the intercept α_0 in equation (4) to

$$\alpha_i = \rho_{i0} + \sum_{j=1}^{S} \rho_{ij} d_j \tag{8}$$

Where d_f is the fth variable of a total number of S variables including household characteristics and a variable for month and year.

Two-stage budgeting design

To obtain unconditional elasticities for alcoholic beverages, which are more suitable for policy recommendations than conditional elasticities (Abdulai 2002, Klonaris and Hallam 2003), we adopt a two-stage budgeting system, assuming separability of the utility function (for further explanations see e.g., Moschini et al. 1994 or Brehe 2007)[a]. First, we assume that the household allocates its budget on the three aggregated product groups at stage 1: food, beverages, and other products and services. At stage 2, the total expenditure for beverages is allocated to the disaggregated product categories: wine, beer, spirits, coffee, tea, cocoa beverages, mineral water, non-alcoholic soft drinks, and fruit and vegetables juices. We estimate income and price response at both stages, while the elasticity at stage 2 is conditional with respect to total expenditure for beverages. Unconditional elasticities for stage 2 could be obtained using the elasticity estimates at stage 1 (see Section Elasticity estimates).

Method and data

Data description and price computation

Summary statistics and definition of the household segments

For our study, we used data collected by the Swiss Federal Statistical Office from 2000 to 2009 for the Swiss household expenditure survey , a national representative survey that consists of cross-sectional data with a periodicity of one month. Almost 3,000 households participate in the survey every year. We can base our estimates on more than 34,000 households over a period of 10 years. Households are chosen randomly from the register or private telephone lines (see SFSO 2011 for details). The sample is stratified with respect to the seven major regions in Switzerland and the distribution of households is calibrated to the distribution of households in the Swiss population (SFSO 2011). For every household, data on expenditure and quantity bought (in liters per month) for all products and services, household income, and detailed information on household characteristics were gathered. To increase the explanatory power of the model and to test whether household characteristics have an influence on alcohol demand, we introduced the following criteria into the model: household size in terms of adult equivalents, a dummy variable for the presence of young children (<5 years), the age of the household's reference person, and a dummy variable for whether the household's reference person has a university degree. Household size is calculated following the OECD-modified equivalence scale (Hagenaars et al. 1994). Following Angulo et al. (2001) we expect a decreasing expenditure share as the number of equivalence increases and the household's reference person's age decreases. With respect to education and the presence of young children there are either contradictory or no clear findings in the literature (see e.g., Van Oers et al. 1999; Reavley et al. 2011; Rice et al. 1998).

The sample is divided into three segments: light drinking households, moderate drinking households, and heavy drinking households. We provide elasticities for the entire sample as well as for the three household segments. The classification is constructed based on the recommendations for Switzerland in Annaheim and Gmel (2004). These are the official recommendations for Switzerland from the Swiss Institute for the Prevention of Alcohol and Drug Problems. Table 1 summarizes the class boundaries in pure alcohol per person and day.

According to FOPH (2014b), we assume the following levels of pure alcohol for each beverage category: 86.9 g pure alcohol per liter for wine, 37.9 g pure alcohol per liter for beer, and 316.0 g pure alcohol per liter for spirits. We calculated the amount of pure alcohol bought per month for every household based on data from the household expenditure survey. We divided the total amount of pure alcohol by the adult equivalent for every household[b] and classified the households according to the class boundaries (Table 1) based on Annaheim and Gmel's recommendations (2004).

Tables 2, 3 and 4 present summary statistics for stage 1, stage 2, and the household characteristics considered in the model. Furthermore, we provide the share of zero consumption for every product group. Given that some households did not buy beverages during the collection period, the sample for stage 2 declines to 33,364 households.

As Table 2 demonstrates, Swiss households spend only 1.4% of their budget on beverages, with 0.8% on alcoholic beverages and 0.7% on non-alcoholic beverages (Table 3). This is much lower than the European Union average values of about 1.5% and 1.2% of household budgets spent on alcoholic and non-alcoholic beverages, respectively (Eurostat 2013). Within alcoholic beverages, the highest share of the Swiss budget is spent on wine, followed by spirits and then beer. Table 3 clearly delineates the differences among the three household segments. Moderate drinking households spend on average four to five times more on wine, beer, or spirits than light drinking households, whereas heavy drinking households spend about twice as much or more in comparison to moderate drinking households. On average, the light drinking household consumes 49.3 g of pure alcohol per month, the moderate drinking household 864.5 g, and the heavy drinking household consume 1890.0 g.

The share of zero consumption is a point of particular interest. Zero consumption means that the left side of equation (4) (budget share) is left censored at zero in the case of demand systems. Besides a short collection period, which may play a special role in the context of beverages, there are other determinants for zero consumption like income restrictions that force households into a corner solution or forgoing

Table 1 Class boundaries in pure alcohol per person and day

	Pure alcohol (in g per day)
Light drinkers	<20
Moderate drinkers	20-39
Heavy drinkers	>40

Table 2 Summary statistics for the stage 1 model, total number of households and expenditure per month

Total number of households	34,176		
	Mean	Standard deviation	Percentage of zero consumption
Total expenditure (in Swiss Francs)	8,432.70	5,132.24	0.00
Expenditure on food (in Swiss Francs)	616.63	354.78	0.14
Expenditure on beverages (in Swiss Francs)	121.63	186.97	2.38
All other expenditure (in Swiss Francs)	1395.42	4933.60	0.00

preferences for certain products (see e.g., Perali and Chavas 2000; Thiele 2008). On the aggregated level (stage 1), zero consumption is low, but at the stage 2 level, a large proportion of households did not consume at all during the collection period. This has serious implications for the model specification. Censoring of the expenditure data should therefore be considered, in particular at stage 2, to avoid biased parameter estimates.

Quality-adjusted prices

Only a few household expenditure surveys collect price data. In the literature, there are two main methods for retrieving price data from expenditure and quantity variables: using unit values as proxy for market prices, as Abdulai (2002) or Akbay et al. (2007) did by dividing the expenditure by the quantity consumed. The other method of quality-adjusted unit values was originally introduced by Cox and Wohlgenant (1986) and is frequently applied in the literature (e.g., in Lazaridis 2003; Thiele 2010; Zheng and Henneberry 2010). Despite the nomenclature, unit values still have a quality aspect, because of the heterogeneity of products within a given category (e.g., wine). The consumer can select between different quality levels, leading to different unit values among households; differences in unit values can be attributed to quality and price variations (Chung et al. 2005). An increase in the unit value could be induced by a price increase or by a shift in the household's demand for more expensive products within this particular product category. Thus, taking unit values as proxies for market prices could lead to biased parameter estimates. This is particularly true for composite commodities (Chung et al. 2005), such as the beverage product groups used in this study, which contain products of varying qualities. As McKelvey (2011) notes, unit values will be an "error-ridden indicator of prices".

Cox and Wohlgenant (1986) proposed correcting the unit value for every household and product group by regressing proxies for quality variations (e.g., education or house-hold size) on the unit value. This leads to quality-adjusted prices that vary between household. Majumder et al. (2012) criticized this approach, stating that households should face the same price at least in the same regional market and proposed a new method to calculate quality-adjusted prices per region. Aepli and Finger (2013) expanded it by a variable for month and year and adapted it for the Swiss household expenditure survey.

We give a short overview of the method (for further details, see Majumder et al. (2012) and Aepli and Finger (2013)). As noted in Majumder et al. (2012), household characteristics and income are good proxies for quality preferences. For instance,

Table 3 Summary statistics for the stage 2 model, sample sizes (reduced samples)[j] and expenditure per month

	Mean (all households)	Standard deviation (all households)	Percentage of zero consumption (all households)	Mean (only light drinking households)	Standard deviation (only light drinking households)	Mean (only moderate drinking households)	Standard deviation (only moderate drinking households)	Mean (only heavy drinking households)	Standard deviation (only heavy drinking households)
Number of households		33,364			25,699		4,152		3,513
Expenditure on beverages (in Swiss Francs)	124.59	188.25	2.38	79.28	69.85	192.70	128.76	375.54	444.81
Expenditure on alcoholic beverages (in Swiss Francs)	67.57	175.72	32.09	23.93	39.15	131.25	109.74	311.51	435.48
Expenditure on non-alcoholic beverages (in Swiss Francs)	57.02	53.95	1.61	55.35	53.30	61.45	53.57	64.03	58.19
Wine (in Swiss Francs)	48.31	162.57	49.10	15.57	34.37	91.17	114.12	237.16	425.85
Beer (in Swiss Francs)	8.23	20.55	69.52	4.36	11.18	17.23	26.65	25.87	41.73
Spirits, sweet wines, etc. (in Swiss Francs)	11.03	32.44	71.33	4.00	13.47	22.85	34.81	48.48	72.84
Coffee (in Swiss Francs)	14.46	31.21	44.66	13.49	30.03	17.23	31.81	18.32	37.81
Tea (in Swiss Francs)	3.87	8.83	59.92	3.91	9.09	3.88	7.76	3.57	8.05
Cocoa beverages (in Swiss Francs)	1.92	6.36	85.05	2.03	6.61	1.64	5.45	1.48	5.45
Mineral water (in Swiss Francs)	10.89	19.33	39.88	10.33	19.17	12.46	19.28	13.10	20.26
Non-alcoholic soft drinks (in Swiss Francs)	15.87	25.01	32.62	15.67	24.64	16.17	25.77	17.02	26.68
Fruit juices and vegetable juices (in Swiss Francs)	10.01	19.81	38.50	9.92	20.65	10.06	15.62	10.54	17.89

Table 4 Summary statistics for the household characteristics

Household characteristics	Mean	Standard deviation
Household size in adult equivalents	1.63	0.52
Young children (<5 years, yes or no)	0.14	0.35
Age of the household's reference person	49.69	15.22
University degree with respect to the household's reference person (yes or no)	0.13	1.27

households with less income tend to choose less expensive goods within a product group. A separate regression is run for each product category i. Following Aepli and Finger (2013), the unit value and the proxy variables are related as follows:

$$v_i^{hlym} - \left(v_i^{lym}\right)_{median} = \alpha_i D_l + \beta_i D_y + \gamma_i D_m + \delta_i x^{hlym} + \eta_i x^{2hlym} + \theta_i e^{hlym} + \omega_i f_i^{hlym}$$
$$+ \sum_{j=1}^{n} b_i Z_{ij}^{hlym} + \varepsilon_i^{hlym}$$

$$(9)$$

where v_i^{hlym} is the unit value paid by household h for item i in its region l, year y, and month m. The deviation of the household's unit value from the median is explained by household characteristics, income, and expenditure variables, which capture the quality effect. The part of the deviation that cannot be explained by quality is captured by the error term ε_i^{hlym}. Furthermore, x denotes income and x^2 the square of income, e is the household total expenditure for beverages, and f is the household total expenditure for food and beverages consumed away from home. Z_{ij} denotes the j th of n household characteristics, which are household size, a binary dummy variable for having children, and a dummy variable for having a university degree. D_l, D_y, and D_m are dummies for region, year, and month, respectively. To manage possible outliers more successfully, we decided to follow the suggestions in Aepli and Finger (2013) and to estimate the equations using a robust M-estimator using only those households that consumed the good.

To obtain regional, monthly, and yearly quality adjusted market prices p_i^{lmy} we add the median unit value v_i^{lym} median to the corresponding median of item i of the estimated residuals from equation (9).

$$\left(p_i^{lmy}\right)_{median} = \left(v_i^{lmy}\right)_{median} + \left(\varepsilon_i^{lmy}\right)_{median} \qquad (10)$$

The prices are then assigned to all households in the sample according to region, month, and year. Due to a low number of households consuming tea and cocoa beverages in the Italian-speaking part of Switzerland in some months, we took the corresponding consumer price index for those product categories, as well as for all product categories at stage 1.

Estimation procedure and elasticity calculation
Censoring
Table 3 shows that the share of zero consumption is relatively high for all beverage product categories at stage 2. This fact has some major consequences on the estimation procedure and must be taken into account to avoid self-selection and resulting biased

parameter estimates. As Deaton (1990) states, deleting the zero-consumption points only allows for estimating conditional effects. The most popular model for coping with censored data is the Tobit approach (Tobin 1958; Amemiya 1984), which is widely used for single-equation demand models. Heien and Wessels (1990) proposed a new approach for equation systems based on Heckman (1979) (for its application see e.g., Dey et al. 2011; Lazaridis 2003). Their approach was criticized by Shonkwiler and Yen (1999) and later by Vermeulen (2001), who showed that the revised procedure leads to inconsistent estimates. As an alternative, Shonkwiler and Yen (1999) proposed another frequently used approach (e.g., in Thiele 2008; Su and Yen 2000; Zheng and Henneberry 2011; for wine in particular, see e.g., Stasi et al. 2011) that is applied in this study.

In the first step, a multivariate probit is estimated (unlike Shonkwiler and Yen (1999), who estimated a univariate probit). We suppose that one household's purchase of a particular beverage product category is not independent of purchase decisions for other beverage product categories due to substitution effects. The model to be estimated is:

$$w_i^* = f\left(x_{ij}\beta_{ij}\right) + \varepsilon_i, \; g_i^* = z_{ik}'\gamma_{ik} + v_i \; i = 1, ..., \; n \; i = 1, ..., \; m \; k = 1, ..., f \tag{11}$$

$$g_i = \begin{cases} 0 \; \text{if } g_i^* \le 0 \\ 1 \; \text{if } g_i^* > 0 \end{cases} \; w_i = g_i w_i^* \, i = 1, ..., \; n \tag{12}$$

where w_i and g_i are the observed dependent variables for the food product groups, g_i^* and w_i^* are the correspondent latent variables, g_i is a binary variable representing the decision of the household to consume or not, and x_{ij} and z'_{ik} are explanatory variables such as income, logarithmic prices, and household characteristics. All these variables are presumed to be important determinants with respect to the decision of the household to consume or not (in this context we refer also to Thiele 2008 and Zheng and Henneberry 2011). The error terms ε_i and v_i are assumed to have a multivariate normal distribution, each with a mean of zero and a variance-covariance matrix V with diagonal elements of 1 and off-diagonal elements of $\rho_{rl} = \rho_{lr}$.

Based on the estimated parameters, we calculate the standard normal cumulative distribution function $\Phi( )$ (cdf) and the standard normal probability density function $\phi( )$ (pdf) for every household and each product category; cdf and pdf are used to correct the budget share equation (4) for a censoring of the budget share variables as follows:

$$w_i = \Phi\left(z_{ik}'\gamma_{ik}\right) f\left(x_{ij}\beta_{ij}\right) + \delta_i \phi\left(z_{ik}'\gamma_{ik}\right) + \varepsilon_i, \; \varepsilon_i \sim MVN(0, \sigma) \, i = 1, ..., \; n \tag{13}$$

Following Zheng and Henneberry (2010), the parameter for the pdf δ_i represents the covariance between the error term in the budget share equation (4) and the error term of the multivariate probit model. Equation (13) is applied only to stage 2. At stage 1, there is no need for the two-step estimation procedure due to the low level of zero consumption. We dropped the equation for expenditure on other products and services as described in Section Theoretical framework of the QUAIDS. For the two-step estimation procedure, the right hand side generally no longer adds up to one, so we estimate the full equation system following Yen et al. (2002) or Ecker and Qaim (2010)[d].

Despite the advantages of Shonkwiler and Yen's (1999) procedure, Tauchmann (2005) draws attention to the problem of heteroscedasticity. By expanding the budget share equation cdf and pdf, heteroscedasticity is implicitly introduced into the model (for further explanations, see Tauchmann 2005). To avoid this problem and to estimate a heteroscedasticity robust covariance matrix, we apply the parametric bootstrap.

Expenditure endogeneity

Assuming weak separability and applying a two-stage budgeting model could lead to problems with respect to the endogeneity of the expenditure variable in the budget share equation, because the expenditure variable could be correlated with the equation errors (Attfield 1985). To account for endogeneity, Blundell and Robin (1999) proposed the augmented regression technique. In a first step, we regress all the price variables, the household characteristics, and the variable for month and year of the budget share equation on the expenditure variable for every product category. On the right hand side, we further include income and its square as instruments as proposed by Bopape (2006). In contrast to other studies, where the equations are estimated using OLS (e.g., Fashogbon and Oni 2013), we use a robust M-estimator to account for possible outliers. Residuals are then included in every budget share equation as additional right hand side variables. Testing the parameter of the residuals for significance allows us to check whether endogeneity is present or not.

Elasticity estimates

Income elasticities for stage 2, where censoring is an issue, are derived using the formula proposed by

$$e_i = \frac{\Phi(z'_{ik}\gamma_{ik})\mu_i}{w_i} + 1 \tag{14}$$

which reduces to $e_i = \frac{\mu_i}{w_i} + 1$ (Banks et al. 1997) for stage 1, where μ_i is the differentiation of equation (4) with respect to $\log m^e$. For the Marshallian price elasticity ε_{ij}^M, we follow Zheng and Henneberry (2010) and calculate the full effect of a price change on demand based on the effect of the budget share equation and the effect of the multivariate probit model for stage 2:

$$\varepsilon_{ij}^M = \frac{\Phi(z'_{ik}\gamma_{ik})\mu_{ij}}{w_i} + \phi * \tau_{ij} * \left(1 - \frac{\delta_i}{w_i}\right) - \delta_{ij} \tag{15}$$

where μ_{ij} is the differentiation of equation (4) with respect to $\log p_j$. [f] τ_{ij} is the estimated parameter for price j with respect to product category i in the multivariate probit model and δ_i is defined as above. δ_{ij} denotes the Kronecker delta, which is equal to one when $i = j$ and otherwise is zero. For stage 1, the formula for the Marshallian price elasticity reduces to $\varepsilon_{ij}^M = \frac{\mu_{ij}}{w_i} - \delta_{ij}$.

The Slutsky equation is applied to get the Hicksian price elasticities:

$$e_{ij}^H = e_{ij}^M + e_i w_j \tag{16}$$

For stage 2, we first estimate conditional elasticities with respect to total expenditures on beverages. Unconditional elasticities are derived following Edgerton (1997). The formula for the income elasticity (unconditional) e_i^{uc} is:

$$e_i{}^{uc} = e_{(r)i}{}^c * e_{(r)}{}^{uc} \tag{17}$$

where $e_{(r)i}^c$ is the expenditure elasticity (conditional) for item i within the r th product group and $e_{(r)}^{uc}$ the income elasticity for the r th product group (unconditional). The unconditional Marshallian price elasticity $\varepsilon_{ij}^{M,\ uc}$ is calculated by:

$$\varepsilon_{ij}^{M,\ uc} = \varepsilon_{(r)ij}^{H,\ c} + e_{(r)i}{}^c * w_{(r)j} \varepsilon_{rr}{}^{M,uc} \tag{18}$$

where $\varepsilon_{(r)ij}^{H,\ c}$ is the conditional Marshallian price elasticity for item i and j, $w_{(r)j}$ the budget share of item j within the r th product category, and $\varepsilon_{rr}^{M,uc}$ the uncompensated Hicksian own-price elasticity of product category r.

The unconditional, Hicksian price elasticity $\varepsilon_{ij}^{H,\ uc}$ is derived analogously using $\varepsilon_{(r)ij}^{H,\ c}$ and the Hicksian own-price elasticity (unconditional) of product category r $\varepsilon_{rr}^{H,uc}$:

$$\varepsilon_{ij}^{H,\ uc} = \varepsilon_{(r)ij}^{H,\ c} + e_{(r)i}{}^c * w_{(r)j} \varepsilon_{rr}{}^{H,uc} \tag{19}$$

Results and discussion

Overall, the variables in the model explain between 0.47% and 0.86% of total variance at stage 1 and between 0.06% and 0.80% of total variance at stage 2. For the sake of brevity, we report the parameter estimates of stages 1 and 2, the coefficients of determination, and X^2 -test statistics for the entire model in the Additional file 1: Appendix. While prices and socio-demographic variables do explain a part of the variance in consumption, the rest is determined by other factors such as lifestyle or psychographic or behavioral aspects, which are not considered further in this study (for determinants of Swiss wine demand see e.g., Brunner and Siegrist 2011). While household size is mostly negatively correlated with budget share, the age of the households' reference person shows a positive correlation which is in line with the findings of previous studies. For the presence of young children and education, we could not show a clear relationship with respect to the budget share. The sign of the parameters varies depending on the household segment.

To verify the model specification with respect to the implementation of household characteristics and the quadratic income term, we carried out Wald-tests for every model. The test statistics reject clearly the null-hypothesis of no household characteristics for all models (Additional file 1: Appendix), which can also be derived from the estimation results for the parameter estimates in Tables 24 to 28 (in the Additional file 1: Appendix). The hypothesis that Engel curves for beverages in Switzerland are often of a nonlinear form is supported by the test results. The QUAIDS specification is therefore superior to the AIDS specification for stage 1 and stage 2, except for the models for light drinking and moderate drinking households (Additional file 1: Appendix). These findings are consistent with other findings for food demand in Switzerland (Abdulai 2002; Aepli and Kuhlgatz 2014). Furthermore, we tested for endogeneity of the expenditure variable in the model. The hypothesis of no endogeneity is clearly rejected for all models and confirms the importance of a correction for endogeneity (Additional file 1: Appendix), which we achieved by applying the augmented regression technique.

Unconditional Hicksian price and income elasticities for beverages are reported in Tables 5, 6, 7, 8, 9, 10, 11 and 12 for all households and the three segments. We provide

Table 5 Unconditional, Marshallian price and income elasticities at stage 2, all households (at sample means)

	Win	Bee	Sp	Cof	Tea	CBe	MWa	NaS	FJVJ	Income
Wine	−0.986 (0.208)***	−0.177 (0.142)	0.046 (0.097)	−0.141 (0.181)	−1.328 (0.325)***	0.353 (0.248)	0.354 (0.203)	−0.256 (0.296)	0.054 (0.135)	1.288 (0.182)***
Beer	0.188 (0.139)	−0.634 (0.242)***	0.100 (0.093)	0.153 (0.112)	0.571 (0.260)**	−1.260 (0.238)***	−1.036 (0.180)***	0.002 (0.109)	0.296 (0.153)*	0.096 (0.085)
Spirits	0.397 (0.146)***	−0.164 (0.072)**	−0.950 (0.043)***	−0.082 (0.122)	0.565 (0.357)	0.519 (0.215)**	0.132 (0.116)	−0.070 (0.209)	0.003 (0.069)	0.918 (0.101)***
Coffee	0.055 (0.088)	0.185 (0.084)**	0.098 (0.077)	−0.786 (0.081)***	−2.436 (0.256)***	−1.508 (0.158)***	−0.320 (0.136)**	−0.126 (0.110)	0.172 (0.079)**	0.350 (0.132)***
Tea	−0.157 (0.114)	0.007 (0.108)	−0.030 (0.089)	−0.143 (0.128)	−4.638 (0.506)***	−2.442 (0.294)***	0.300 (0.218)	−0.045 (0.180)	0.030 (0.097)	0.419 (0.160)***
Cocoa beverages	0.443 (0.242)*	−0.593 (0.219)**	−0.138 (0.100)	0.028 (0.207)	−3.365 (0.679)***	−0.753 (0.484)	1.175 (0.236)***	−0.305 (0.330)	−0.267 (0.167)	1.441 (0.106)***
Mineral water	−0.068 (0.182)	−0.416 (0.115)***	−0.106 (0.066)	0.016 (0.166)	−2.969 (0.273)***	−0.064 (0.229)	−0.186 (0.148)	−0.213 (0.262)	0.058 (0.110)	1.156 (0.107)***
Non-alcoholic soft drinks	−0.011 (0.094)	0.054 (0.076)	0.099 (0.034)***	−0.031 (0.076)	−0.406 (0.148)***	−0.565 (0.124)***	−0.403 (0.079)***	−1.169 (0.122)***	0.226 (0.065)***	0.518 (0.044)***
Fruit juices and vegetables juices	0.033 (0.073)	0.318 (0.109)***	0.164 (0.046)***	−0.253 (0.074)***	−1.933 (0.210)***	−1.279 (0.167)***	−0.745 (0.124)***	−0.053 (0.101)	−0.757 (0.086)***	0.363 (0.057)***

*, ** and *** denote significance at 10%, 5% and 1% level, respectively.

Note: Win: Wine, Bee: Beer, Sp: Spirits, Cof: Coffee, Tea: Tea, CBe: Cocoa beverages, MWa: Mineral Water, NaS: Non-alcoholic soft drinks, FJVJ: Fruit juices and vegetables juices.

Table 6 Unconditional, Hicksian price and income elasticities at stage 2, all households (at sample means)

	Win	Bee	Sp	Cof	Tea	CBe	MWa	NaS	FJVJ	Income
Wine	-0.983 (0.208)***	-0.176 (0.142)	0.046 (0.097)	-0.138 (0.181)	-1.324 (0.325)***	0.356 (0.248)	0.355 (0.203)*	-0.252 (0.295)	0.055 (0.135)	1.288 (0.182)***
Beer	0.188 (0.139)	-0.634 (0.242)***	0.100 (0.093)	0.153 (0.112)	0.571 (0.260)**	-1.260 (0.238)***	-1.036 (0.180)***	0.002 (0.109)	0.296 (0.153)*	0.096 (0.085)
Spirits	0.399 (0.146)***	-0.163 (0.072)**	-0.950 (0.043)***	-0.080 (0.122)	0.567 (0.357)	0.520 (0.215)**	0.133 (0.116)	-0.067 (0.209)	0.004 (0.069)	0.918 (0.101)***
Coffee	0.056 (0.088)	0.185 (0.084)**	0.098 (0.077)	-0.785 (0.081)***	-2.435 (0.256)***	-1.507 (0.158)***	-0.320 (0.136)**	-0.125 (0.110)	0.172 (0.079)**	0.350 (0.132)***
Tea	-0.156 (0.113)	0.008 (0.108)	-0.030 (0.089)	-0.142 (0.128)	-4.637 (0.506)***	-2.442 (0.294)***	0.300 (0.218)	-0.044 (0.180)	0.031 (0.097)	0.419 (0.160)***
Cocoa beverages	0.446 (0.242)*	-0.592 (0.219)***	-0.138 (0.100)	0.031 (0.207)	-3.361 (0.679)***	-0.751 (0.484)	1.176 (0.236)***	-0.300 (0.330)	-0.266 (0.167)	1.441 (0.106)***
Mineral water	-0.066 (0.182)	-0.415 (0.115)***	-0.105 (0.066)	0.018 (0.166)	-2.966 (0.273)***	-0.062 (0.229)	-0.185 (0.148)	-0.209 (0.262)	0.059 (0.109)	1.156 (0.107)***
Non-alcoholic soft drinks	-0.010 (0.094)	0.054 (0.076)	0.099 (0.034)***	-0.030 (0.076)	-0.404 (0.148)***	-0.564 (0.124)***	-0.402 (0.079)***	-1.167 (0.122)***	0.226 (0.065)***	0.518 (0.044)***
Fruit juices and vegetables juices	0.034 (0.073)	0.318 (0.109)***	0.164 (0.046)***	-0.252 (0.074)***	-1.932 (0.210)***	-1.278 (0.167)***	-0.744 (0.124)***	-0.052 (0.101)	-0.757 (0.086)***	0.363 (0.057)***

*, ** and *** denote significance at 10%, 5% and 1% level, respectively.

Note: Win: Wine, Bee: Beer, Sp: Spirits, Cof: Coffee, Tea: Tea, CBe: Cocoa beverages, MWa: Mineral Water, NaS: Non-alcoholic soft drinks, FJVJ: Fruit juices and vegetables juices.

Table 7 Unconditional, Marshallian price and income elasticities at stage 2, light drinking households (at sample means)

	Win	Bee	Sp	Cof	Tea	CBe	MWa	NaS	FJVJ	Income
Wine	−1.275 (0.247)***	−0.144 (0.141)	0.038 (0.203)	−0.156 (0.161)	−0.801 (1.050)	0.172 (0.718)	0.450 (0.348)	−0.484 (0.177)***	0.003 (0.595)	0.620 (0.366)*
Beer	0.679 (0.707)	−1.216 (0.457)***	0.000 (0.265)	0.550 (0.226)**	−1.892 (2.355)	−2.940 (1.576)*	−1.056 (1.363)	0.480 (0.545)	1.082 (0.874)	0.469 (0.534)
Spirits	0.363 (0.400)	−0.011 (0.151)	−0.859 (0.220)***	0.109 (0.152)	2.874 (1.163)**	0.317 (1.276)	−0.760 (0.730)	0.429 (0.235)*	0.417 (0.703)	0.666 (0.408)
Coffee	0.365 (0.348)	0.165 (0.151)	0.036 (0.163)	−0.729 (0.134)***	−3.209 (1.311)**	−1.825 (0.721)**	−0.431 (0.667)	0.192 (0.240)	0.708 (0.527)	0.569 (0.307)*
Tea	−1.000 (0.580)*	−0.438 (0.245)*	−0.070 (0.430)	−0.606 (0.297)**	−4.656 (2.844)	−0.586 (1.466)	2.528 (1.147)**	−0.721 (0.343)**	−1.146 (1.403)	0.875 (0.845)
Cocoa beverages	0.635 (0.564)	−0.332 (0.627)	0.014 (0.403)	−0.006 (0.446)	−4.249 (1.628)***	−1.244 (1.244)	−0.807 (0.915)	0.375 (0.610)	−0.574 (0.825)	1.489 (0.668)**
Mineral water	−0.193 (0.831)	−0.854 (0.670)	−0.099 (0.316)	−0.710 (0.445)	−6.929 (2.611)***	0.722 (1.186)	0.490 (1.599)	−0.193 (0.653)	−1.239 (0.819)	2.068 (0.403)***
Non-alcoholic soft drinks	−0.282 (0.255)	0.100 (0.230)	−0.011 (0.100)	0.074 (0.154)	−0.168 (0.348)	−0.422 (0.339)	0.141 (0.437)	−1.062 (0.256)***	0.384 (0.129)***	0.464 (0.109)***
Fruit juices and vegetables juices	0.171 (0.275)	0.388 (0.273)	0.111 (0.163)	0.025 (0.138)	−0.812 (1.077)	−1.558 (0.458)***	−0.262 (0.753)	0.025 (0.251)	−0.206 (0.465)	0.220 (0.248)

*, ** and *** denote significance at 10%, 5% and 1% level, respectively.
Note: Win: Wine, Bee: Beer, Sp: Spirits, Cof: Coffee, Tea: Tea, CBe: Cocoa beverages, MWa: Mineral Water, NaS: Non-alcoholic soft drinks, FJVJ: Fruit juices and vegetables juices.

Table 8 Unconditional, Hicksian price and income elasticities at stage 2, light drinking households (at sample means)

	Win	Bee	Sp	Cof	Tea	CBe	MWa	NaS	FJVJ	Income
Wine	-1.273 (0.246)***	-0.143 (0.141)	0.039 (0.203)	-0.155 (0.161)	-0.799 (1.049)	0.174 (0.718)	0.450 (0.348)	-0.483 (0.176)***	0.004 (0.595)	0.620 (0.366)*
Beer	0.680 (0.707)	-1.216 (0.457)***	0.001 (0.265)	0.551 (0.225)*	-1.890 (2.355)	-2.938 (1.576)*	-1.056 (1.363)	0.480 (0.545)	1.082 (0.874)	0.469 (0.534)
Spirits	0.365 (0.400)	-0.010 (0.151)	-0.858 (0.220)***	0.111 (0.152)	2.877 (1.163)**	0.319 (1.276)	-0.760 (0.730)	0.430 (0.235)*	0.417 (0.703)	0.666 (0.408)
Coffee	0.367 (0.347)	0.165 (0.151)	0.036 (0.163)	-0.728 (0.133)***	-3.207 (1.311)**	-1.823 (0.721)**	-0.431 (0.667)	0.193 (0.240)	0.708 (0.527)	0.569 (0.307)*
Tea	-0.998 (0.579)*	-0.437 (0.245)*	-0.070 (0.430)	-0.604 (0.296)**	-4.653 (2.843)	-0.584 (1.466)	2.529 (1.147)**	-0.720 (0.342)**	-1.146 (1.403)	0.875 (0.845)
Cocoa beverages	0.639 (0.564)	-0.330 (0.627)	0.014 (0.403)	-0.003 (0.446)	-4.244 (1.627)***	-1.240 (1.244)	-0.806 (0.915)	0.377 (0.609)	-0.573 (0.825)	1.489 (0.668)**
Mineral water	-0.188 (0.831)	-0.852 (0.670)	-0.098 (0.316)	-0.705 (0.445)	-6.923 (2.611)***	0.727 (1.186)	0.491 (1.599)	-0.189 (0.653)	-1.238 (0.819)	2.068 (0.403)***
Non-alcoholic soft drinks	-0.280 (0.254)	0.100 (0.230)	-0.010 (0.100)	0.075 (0.154)	-0.166 (0.348)	-0.421 (0.339)	0.141 (0.437)	-1.061 (0.256)***	0.384 (0.129)***	0.464 (0.109)***
Fruit juices and vegetables juices	0.171 (0.275)	0.388 (0.273)	0.112 (0.163)	0.026 (0.137)	-0.812 (1.077)	-1.558 (0.458)***	-0.262 (0.753)	0.026 (0.251)	-0.206 (0.465)	0.220 (0.248)

*, ** and *** denote significance at 10%, 5% and 1% level, respectively.
Note: Win: Wine, Bee: Beer, Sp: Spirits, Cof: Coffee, Tea: Tea, CBe: Cocoa beverages, MWa: Mineral Water, NaS: Non-alcoholic soft drinks, FJVJ: Fruit juices and vegetables juices.

Table 9 Unconditional, Marshallian price and income elasticities at stage 2, moderate drinking households (at sample means)

	Win	Bee	Sp	Cof	Tea	CBe	MWa	NaS	FJVJ	Income
Wine	−0.665 (1.419)	0.194 (1.155)	0.236 (0.606)	−0.029 (0.554)	0.186 (0.614)	−0.183 (0.548)	−0.362 (0.450)	−0.562 (0.563)	−0.088 (0.232)	1.077 (0.355) ***
Beer	0.639 (1.230)	−0.825 (1.186)	0.365 (0.535)	−0.049 (0.699)	1.096 (1.289)	−0.893 (0.554)	−0.301 (0.450)	−0.317 (0.704)	−0.137 (0.266)	0.877 (0.867)
Spirits	0.305 (0.954)	0.166 (0.453)	−0.799 (0.351)**	−0.093 (0.249)	−0.877 (0.514)*	−0.479 (0.304)	−0.143 (0.310)	−0.423 (0.390)	−0.144 (0.145)	0.636 (0.355) *
Coffee	0.034 (1.414)	0.015 (1.579)	0.077 (0.686)	−0.828 (0.611)	−1.718 (1.648)	−0.250 (1.180)	0.173 (0.473)	−0.096 (0.479)	0.105 (0.361)	0.058 (0.436)
Tea	0.264 (2.398)	0.005 (1.583)	0.043 (0.960)	−0.237 (0.757)	−1.360 (1.488)	−1.367 (1.008)	0.524 (0.685)	0.069 (0.714)	−0.170 (0.403)	0.241 (0.613)
Cocoa beverages	0.337 (2.472)	−1.143 (2.480)	−0.097 (1.039)	−0.063 (1.441)	−2.195 (3.197)	1.858 (3.379)	−0.087 (1.077)	−0.129 (1.496)	0.389 (0.857)	−0.186 (1.539)
Mineral water	−0.293 (1.706)	−0.021 (0.873)	−0.033 (0.561)	−0.212 (0.523)	−0.716 (0.757)	−0.136 (0.620)	−0.512 (1.234)	−0.211 (0.704)	0.316 (0.563)	0.632 (0.509)
Non-alcoholic soft drinks	0.145 (1.497)	−0.034 (0.870)	0.009 (0.553)	−0.004 (0.435)	−0.724 (0.605)	−0.053 (0.577)	0.023 (0.547)	−1.141 (0.550)**	−0.051 (0.286)	0.489 (0.420)
Fruit juices and vegetables juices	0.362 (0.973)	−0.058 (0.937)	0.007 (0.384)	−0.041 (0.452)	−2.267 (0.857)***	0.265 (0.457)	0.175 (0.638)	−0.419 (0.417)	−0.947 (0.356)***	0.503 (0.329)

*, ** and *** denote significance at 10%, 5% and 1% level, respectively.
Note: Win: Wine, Bee: Beer, Sp: Spirits, Cof: Coffee, Tea: Tea, CBe: Cocoa beverages, MWa: Mineral Water, NaS: Non-alcoholic soft drinks, FJVJ: Fruit juices and vegetables juices.

Table 10 Unconditional, Hicksian price and income elasticities at stage 2, moderate drinking households (at sample means)

	Win	Bee	Sp	Cof	Tea	CBe	MWa	NaS	FJVJ	Income
Wine	−0.663 (1.419)	0.194 (1.155)	0.236 (0.606)	−0.028 (0.554)	0.188 (0.614)	−0.182 (0.548)	−0.360 (0.450)	−0.556 (0.563)	−0.087 (0.232)	1.077 (0.355)***
Beer	0.640 (1.230)	−0.825 (1.186)	0.365 (0.535)	−0.048 (0.699)	1.098 (1.289)	−0.892 (0.554)	−0.299 (0.449)	−0.312 (0.701)	−0.135 (0.265)	0.877 (0.867)
Spirits	0.306 (0.954)	0.166 (0.453)	−0.799 (0.351)**	−0.093 (0.249)	−0.876 (0.514)*	−0.479 (0.304)	−0.141 (0.310)	−0.420 (0.389)	−0.143 (0.145)	0.636 (0.355)*
Coffee	0.034 (1.414)	0.015 (1.579)	0.077 (0.686)	−0.828 (0.611)	−1.717 (1.648)	−0.250 (1.180)	0.173 (0.473)	−0.095 (0.478)	0.105 (0.361)	0.058 (0.436)
Tea	0.265 (2.398)	0.006 (1.583)	0.043 (0.960)	−0.237 (0.757)	−1.360 (1.488)	−1.367 (1.008)	0.525 (0.685)	0.070 (0.712)	−0.170 (0.402)	0.241 (0.613)
Cocoa beverages	0.336 (2.472)	−1.143 (2.480)	−0.097 (1.039)	−0.063 (1.441)	−2.195 (3.197)	1.857 (3.378)	−0.088 (1.076)	−0.130 (1.491)	0.389 (0.856)	−0.186 (1.539)
Mineral water	−0.292 (1.706)	−0.021 (0.873)	−0.033 (0.561)	−0.212 (0.523)	−0.715 (0.757)	−0.135 (0.620)	−0.511 (1.233)	−0.207 (0.703)	0.317 (0.563)	0.632 (0.509)
Non-alcoholic soft drinks	0.146 (1.497)	−0.034 (0.870)	0.009 (0.553)	−0.003 (0.435)	−0.723 (0.604)	−0.052 (0.577)	0.024 (0.547)	−1.139 (0.549)**	−0.050 (0.286)	0.489 (0.420)
Fruit juices and vegetables juices	0.362 (0.973)	−0.057 (0.937)	0.007 (0.384)	−0.040 (0.452)	−2.266 (0.857)***	0.265 (0.457)	0.176 (0.638)	−0.416 (0.416)	−0.947 (0.356)***	0.503 (0.329)

*, ** and *** denote significance at 10%, 5% and 1% level, respectively.
Note: Win: Wine, Bee: Beer, Sp: Spirits, Cof: Coffee, Tea: Tea, CBe: Cocoa beverages, MWa: Mineral Water, NaS: Non-alcoholic soft drinks, FJVJ: Fruit juices and vegetables juices.

Table 11 Unconditional, Marshallian price and income elasticities at stage 2, heavy drinking households (at sample means)

	Win	Bee	Sp	Cof	Tea	CBe	MWa	NaS	FJVJ	Income
Wine	-0.519 (0.319)	0.573 (0.557)	0.251 (0.204)	0.178 (0.119)	1.533 (1.075)	-1.456 (0.803)*	-0.641 (0.295)**	-0.673 (0.516)	0.543 (0.325)*	1.009 (0.172)***
Beer	0.800 (1.210)	-0.244 (2.026)	0.289 (0.671)	0.498 (0.604)	4.057 (5.212)	-3.259 (4.122)	-1.306 (1.427)	-0.684 (0.987)	1.419 (1.960)	1.412 (0.768)**
Spirits	0.336 (0.512)	-0.684 (0.897)	-0.919 (0.315)***	-0.260 (0.221)	-3.335 (1.771)*	1.735 (1.339)	0.494 (0.451)	-0.284 (0.372)	-1.095 (0.566)*	0.357 (0.240)
Coffee	0.066 (0.451)	-0.451 (0.819)	0.237 (0.275)	-0.997 (0.318)***	-3.220 (2.295)	1.173 (1.721)	0.408 (0.586)	-0.296 (0.402)	-0.898 (0.855)	0.488 (0.338)
Tea	0.230 (0.830)	-1.307 (1.232)	0.199 (0.522)	-0.319 (0.416)	-2.582 (3.903)	3.393 (2.865)	1.149 (0.799)	-0.631 (0.674)	-0.859 (1.050)	0.258 (0.503)
Cocoa beverages	0.131 (0.873)	-0.429 (1.159)	0.100 (0.569)	-0.285 (0.655)	-6.955 (5.165)	-1.639 (3.760)	0.055 (1.388)	-0.274 (0.790)	-0.864 (2.016)	0.426 (0.634)
Mineral water	0.079 (0.468)	-0.442 (0.783)	-0.055 (0.343)	-0.374 (0.457)	-2.208 (2.190)	1.556 (1.655)	-0.686 (0.775)	-0.591 (0.491)	0.007 (1.110)	0.398 (0.350)
Non-alcoholic soft drinks	0.483 (0.545)	-0.193 (1.027)	0.069 (0.333)	-0.255 (0.329)	-2.633 (2.429)	1.908 (1.947)	0.308 (0.649)	-1.702 (0.421)***	-0.931 (0.951)	0.389 (0.378)
Fruit juices and vegetables juices	0.233 (0.826)	-0.969 (1.339)	-0.082 (0.410)	-0.691 (0.622)	-2.330 (3.855)	1.553 (3.069)	1.517 (1.340)	-0.160 (0.623)	-2.154 (1.735)	0.395 (0.601)

*, ** and *** denote significance at 10%, 5% and 1% level, respectively.
Note: Win: Wine, Bee: Beer, Sp: Spirits, Cof: Coffee, Tea: Tea, CBe: Cocoa beverages, MWa: Mineral Water, NaS: Non-alcoholic soft drinks, FJVJ: Fruit juices and vegetables juices.

Table 12 Unconditional, Hicksian price and income elasticities at stage 2, heavy drinking households (at sample means)

	Win	Bee	Sp	Cof	Tea	CBe	MWa	NaS	FJVJ	Income
Wine	−0.518 (0.319)	0.574 (0.557)	0.251 (0.204)	0.179 (0.119)	1.535 (1.075)	−1.455 (0.803)**	−0.639 (0.295)**	−0.666 (0.516)	0.545 (0.325)*	1.009 (0.172)***
Beer	0.801 (1.210)	−0.244 (2.026)	0.289 (0.671)	0.499 (0.604)	4.059 (5.212)	−3.258 (4.122)	−1.303 (1.426)	−0.674 (0.984)	1.420 (1.960)	1.412 (0.768)**
Spirits	0.337 (0.512)	−0.684 (0.897)	−0.919 (0.315) ***	−0.260 (0.221)	−3.334 (1.771)*	1.735 (1.339)	0.495 (0.451)	−0.281 (0.371)	−1.094 (0.566)*	0.357 (0.240)
Coffee	0.067 (0.451)	−0.451 (0.819)	0.237 (0.275)	−0.996 (0.318)***	−3.219 (2.295)	1.173 (1.721)	0.409 (0.586)	−0.292 (0.400)	−0.897 (0.855)	0.488 (0.338)
Tea	0.231 (0.830)	−1.307 (1.232)	0.199 (0.522)	−0.319 (0.416)	−2.582 (3.903)	3.393 (2.865)	1.150 (0.799)	−0.629 (0.671)	−0.859 (1.050)	0.258 (0.503)
Cocoa beverages	0.131 (0.873)	−0.429 (1.159)	0.100 (0.569)	−0.284 (0.655)	−6.954 (5.165)	−1.639 (3.760)	0.056 (1.388)	−0.271 (0.787)	−0.863 (2.016)	0.426 (0.634)
Mineral water	0.079 (0.468)	−0.442 (0.783)	−0.055 (0.343)	−0.373 (0.457)	−2.207 (2.190)	1.556 (1.655)	−0.685 (0.775)	−0.588 (0.489)	0.007 (1.110)	0.398 (0.350)
Non-alcoholic soft drinks	0.484 (0.545)	−0.193 (1.027)	0.069 (0.333)	−0.255 (0.329)	−2.633 (2.429)	1.908 (1.947)	0.309 (0.649)	−1.699 (0.419)***	−0.931 (0.951)	0.389 (0.378)
Fruit juices and vegetables juices	0.234 (0.826)	−0.969 (1.339)	−0.082 (0.410)	−0.691 (0.622)	−2.330 (3.855)	1.553 (3.069)	1.518 (1.340)	−0.157 (0.620)	−2.154 (1.735)	0.395 (0.601)

*, ** and *** denote significance at 10%, 5% and 1% level, respectively.

Note: Win: Wine, Bee: Beer, Sp: Spirits, Cof: Coffee, Tea: Tea, CBe: Cocoa beverages, MWa: Mineral Water, NaS: Non-alcoholic soft drinks, FJVJ: Fruit juices and vegetables juices.

standard errors and the significance level, which are computed by the delta method discussed in Oehlert (1992). We provide the conditional elasticities for stage 2 in the Additional file 1: Appendix. The elasticities for stage 1 can be obtained from the author on request.

General price and income responses for beverages and for the three household segments are discussed in terms of elasticities. Due to the cross-sectional structure of the data set, elasticities should be interpreted as short-term reactions of households to price and income changes.

Income elasticities for alcoholic beverages are all positive in the model for all households. Wine is found to be a luxury good and spirits and beer are found to be necessity goods, though spirits is very closed to the luxury goods. The findings for beer are in line with previous studies (see e.g. Nelson 1997 and Selvanathan and Selvanathan 2005). With respect to wine and spirits the findings in the literature are partially contradictory. In most studies wine and spirits are found to be luxury goods or have an elasticity which is smaller but close to one (Fogarty 2008). Therefore our findings are in the range of variability of the results of previous studies.

Looking at the income elasticities of the three segments, alcoholic beverages are a necessity good for light drinking households, as well as for moderate drinking households, except for wine. For heavy drinking households, beer and wine are luxury goods, while spirits are necessity goods[g]. Light drinking households are clearly less income-elastic for wine and beer than moderate and heavy drinking households, and heavy drinking households are much more income-elastic with respect to beer than moderate drinking households.

Almost all own-price elasticities for alcoholic and non-alcoholic beverages are negative for the three household segments as well as for the model with all households. Therefore, the negativity condition is mostly fulfilled. For wine and beer, our findings show a clear decrease in the magnitude from light drinking households to heavy drinking households. Looking at the Hicksian demand elasticities, an additional 1% decrease in beer price would result in an increase of the light drinking household demand by 1.21%, while moderate drinking households would increase their consumption by 0.83% and heavy drinking households even less by 0.24%. This shows that light drinking households are relatively price elastic while heavy drinking households are inelastic. Beer is of special interest because it is the favorite beverage for young people in Switzerland. A recently study published by Dey et al. (2013) shows that those who practice binge drinking or other risky drinking behavior in Switzerland have a high penchant for beer. The authors (Dey et al. 2013) suggest the relatively low price of beer in comparison to other alcoholic beverages explains these findings. The price argument likely also applies to middle-aged or older people in terms of risky drinking behavior and could be a reason why own-price elasticities for spirits, which are more expensive than beer, are almost constant between the three segments with a range from 0.83 and 0.92[h]. These results are in contrast to Kuo et al. (2003), who found that moderate drinkers show relatively higher price responses than heavy drinkers in relation to spirits.

Looking at the model for all households, non-alcoholic beverages are mostly substitutes for wine and beer. In the case of spirits, tea, cocoa beverages, and mineral water are slight complements, while coffee, non-alcoholic soft drinks, and fruit and vegetables juices are substitutes. With respect to the substitution effects between alcoholic and non-alcoholic beverages among the three households segments, we did not find a clear structure. There are substitutes as well as complementary goods depending on the household segment.

The own-price elasticities for non-alcoholic beverages for the three segments are all negative except for mineral water for light drinking households and cocoa beverages for moderate drinking households. The magnitude of the Hicksian own-price elasticities shows a tendency for higher price sensitivity from light to heavy drinking households, indicating that heavy drinking households are price-sensitive to non-alcoholic beverages in contrast to alcoholic beverages.

Conclusions and policy implications

This paper reports on the income and price elasticities for different alcoholic and non-alcoholic beverage product categories with respect to light, moderate, and heavy drinking households in Switzerland. It is based on microdata of the Swiss household expenditure survey from 2000 to 2009, containing data from more than 34,000 households. We applied a two-stage quadratic almost ideal demand system, correcting for the high share of zero consumption for some product categories at stage 2 and for endogeneity of the expenditure variable. Missing price data were received by adjusting unit values for quality as recently proposed by Aepli and Finger (2013) and Majumder et al. (2012). We tested the model specification with respect to household characteristics as well as the quadratic income term, which distinguishes the QUAIDS from the AIDS. Testing for the overall significance of the household characteristics and the quadratic income terms confirms the model specification. With respect to wine and beer, moderate and heavy drinking households are less price-sensitive than light drinking households are, while for spirits we did not find a difference among the three household segments.

Our findings have major policy implications with respect to the alcohol tax. In general, the higher the negative effects of alcohol consumption and the lower the elasticity, the higher the tax should be set and vice versa. This only holds if we assume a constant elasticity function. By dividing the sample into three segments, we have shown that this does not hold for Swiss households and heavy drinking households are less price-sensitive, in particular with respect to wine and beer, than moderate or light drinking households. To fix the optimal level of a tax on alcohol, the different responses of households to price changes should be considered. As already noted by Manning et al. (1995), the optimal level for an alcohol tax is a trade-off between economic, public health, and social gains due to a reduction in consumption of heavy drinkers and the possible adverse social or even health effects[i] on light or moderate drinkers due to the additional tax burden. Our findings clearly show that the assumption of a constant elasticity function with respect to the three defined segments is violated and that households do not respond similarly to price changes in wine and beer. A tax on those products will therefore lead to a decrease in consumption, especially in light drinking households, and to a lesser extent in moderate drinking households. The effect on heavy drinking households is minimal. From a social and health perspective, this may not be a desirable development. Therefore, before implementing a new tax on alcoholic beverages in Switzerland or in other European countries, the social and health externality costs and the economic effects of a rather small decrease in alcohol consumption among heavy drinking households must be weighed against possible adverse health or economic consequences of a sharp decline in light or moderate drinking households. To a certain extent, our findings for Switzerland can be transferred to other high-income countries and contribute to a differentiated discussion on alcohol tax.

In addition to estimating of the negative external effects of alcohol consumption, a topic for further research will be the estimation of elasticities for different household types with respect to the age of the households' reference person. As Kuo et al. (2003) has already noted, the price response for spirits depends on age. While younger or middle-aged people react more to price changes in spirits, older people are quite inelastic. Our findings show that households react differently to spirits than to wine and beer. Therefore, the findings of Kuo et al. (2003) will probably not hold for wine and beer and a further study should shed valuable light on this question.

Another point of interest is the substitution within a product category, such as shifting from more expensive alcoholic beverages to cheaper ones of lower quality due to a higher tax. Although we were not able to show this effect in our analysis, it is possible that people would buy cheaper products which are likely more risky. Gruenewald et al. (2006) show that consumers respond to price increases by altering their consumption, varying their brand choices as well as substituting between quality classes. This finding has consequences for a tax increase. For example, a tax increase on only high quality alcohol beverages could lead to substitution towards lower quality products and operate as a positive income effect, probably associated with higher consumption of alcoholic beverages in total (Gruenewald et al. 2006). This possibility illustrates the need for research focused on quality substitution.

From a methodological perspective, further research should be conducted on the estimation of price and income for different household segments within a single QUAIDS by introducing cross-terms into the budget share equation, to estimate possible interaction effects between household types and price or income parameters.

Endnotes

[a]After Klonaris and Hallam (2003) conditional elasticities contain only direct effects on demand in comparison to unconditional elasticities, which contain direct and indirect effects. The latter are therefore more relevant in welfare analysis and for policy purposes (Klonaris and Hallam 2003). The reason lies in multistage budgeting; a change in the price for one beverage category at the second stage within beverages (first stage) has a direct effect on the demand of all beverage categories (second stage) but has also an effect on the price index for beverages (first stage) and therefore will influence the allocation at the first stage (beverages and other commodity groups). This could have a further indirect effect on the demand for beverage categories at the second stage.

[b]Adult equivalent is a better proxy for the number of alcohol consuming people in a household than the number of adults because of the relatively high alcohol consumption among people 15 years old or younger in Switzerland (Annaheim and Gmel 2004).

[c]Shonkwiler and Yen's (1999) two-step estimation procedure of does not allow for the incorporation of the adding-up restriction into the model.

[d]$\mu_i \equiv \frac{\partial w_i}{\partial \log m} = \beta_i + \frac{2\lambda_i}{b(p)} \left\{ \log\left[\frac{m}{a(p)}\right] \right\}$

[e]$\mu_{ij} \equiv \frac{\partial w_i}{\partial \log p_j} = \gamma_{ij} - \mu_i \left(\alpha_j + \sum_k \gamma_{jk} \log P_k \right) - \frac{\lambda_i \beta_j}{b(p)} \left\{ \log\left[\frac{m}{a(p)}\right] \right\}^2$

[f]A good is called a luxury good if demand increases more than proportionally as income rises ($e_i > 1$). If demand increases less than proportionally, the good is called a necessity good ($e_i > 1$).

[g]The quality adjusted price for beer (3.00 CHF/Liter) is much lower than the price for wine (10.20 CHF/Liter) or spirits (15.18 CHF/Liter). This supports the interpretation proposed by Dey et al. (2013) that the low price for beer plays an important role with respect to drinking behavior.

[h]The health effects depend on the assumptions made with respect to low or moderate alcohol consumption.

[i]The reduction in sample size in comparison to Table 1 arises due to the deletion of households without consumption of beverages during the data collection period.

Additional file

Additional file 1: Appendix. **Table S1.** Wald tests for household characteristics. **Table S2.** Wald tests for the quadratic term in the QUAIDS. **Table S3.** Wald tests for endogeneity of the expenditure variable. **Table S4.** Conditional, Marshallian price and income elasticities at stage 2, all households (at sample means). **Table S5.** Conditional, Hicksian price and income elasticities at stage 2, all households (at sample means). **Table S6.** Conditional, Marshallian price and income elasticities at stage 2, light drinking household (at sample means). **Table S7.** Conditional, Hicksian price and income elasticities at stage 2, light drinking household (at sample means). **Table S8.** Conditional, Marshallian price and income elasticities at stage 2, moderate drinking household (at sample means). **Table S9.** Conditional, Hicksian price and income elasticities at stage 2, moderate drinking household (at sample means). **Table S10.** Conditional, Marshallian price and income elasticities at stage 2, heavy drinking household (at sample means). **Table S11.** Conditional, Hicksian price and income elasticities at stage 2, heavy drinking household (at sample means). **Table S12.** Parameter estimates of the budget share equation, Stage 1. **Table S13.** Parameter estimates of the budget share equation, Stage 2, all households. **Table S14.** Parameter estimates of the budget share equation, Stage 2, light drinking household. **Table S15.** Parameter estimates of the budget share equation, Stage 2, moderate drinking households. **Table S16.** Parameter estimates of the budget share equation, Stage 2, heavy drinking households.

Competing interests
The authors declare no competing interests.

Authors' contributions
MA prepared the data, constructed the model, performed the statistical analysis and drafted the manuscript.

Acknowledgements
This research was conducted with support from the Swiss Federal Office for Agriculture within the project "Food demand analysis in Switzerland".
We thank Robert Finger, University of Bonn, for his helpful comments on a previous version.

References
Abdulai A (2002) Household demand for food in Switzerland: A quadratic almost ideal demand system. Swiss J Econ Stat 138(1):1–18
Aepli M, Finger R (2013) Determinants of sheep and goat meat consumption in Switzerland. Agri Food Econ, doi: 10.1186/2193-7532-1-11
Aepli M, Kuhlgatz C (2014) Meat and milk demand elasticities for Switzerland: A three stage budgeting Quadratic Almost Ideal Demand System. Submitted
Akbay C, Boz I, Chern WS (2007) Household food consumption in Turkey. Eur Review Agric Econ 34(2):209–231
Amemiya T (1984) Tobit models: A survey. J Econometrics 24(1–2):3–61
Angulo AM, Gil JM, Gracia A (2001) The demand for alcoholic beverages in Spain. Agr Econ 26(1):71–83
Annaheim B, Gmel G (2004) Alkoholkonsum in der Schweiz. Scientific study, schweizerische Fachstelle für Alkohol- und andere Drogenprobleme, Lausanne.
Attfield CLF (1985) Homogeneity and endogeneity in systems of demand equations. J Econometrics 27(2):197–209
Banks J, Blundell R, Lewbel A (1997) Quadratic Engel curves and consumer demand. Rev Econ Stat 79(4):527–539
Barnett WA, Kanyama IK (2013) Time-varying parameters in the almost ideal demand system and the Rotterdam model: Will the best specification please stand up? Appl Econ 45(29):4169–4183
Barnett WA, Seck O (2008) Rotterdam model versus Almost Ideal Demand System: Will the best specification please stand up? J Appl Econom 23(6):795–824
Barten AP (1964) Consumer demand functions under conditions of almost additive preferences. Econometrica 32(1–2):1–38

Blackorby CH, Boyce R, Russell RR (1978) Estimation of demand systems generated by the Gorman polar form: A generalization of the S-Branch utility tree. Econometrica 46(2):345–363

Blundell R, Robin JM (1999) Estimation in large and disaggregated demand systems: An estimator for conditionally linear systems. J Appl Econom 14(3):209–232

Bopape L (2006) The influence of demand model selection on household welfare estimates: An application to South African food expenditures. In: Dissertation. Michigan State University

Bopape L, Myers R (2007) Analysis of household food demand in South Africa: Model selection, expenditure endogeneity, and the influence of socio-demographic effects. Paper presented at the African econometrics society annual conference, Cape Town, South Africa

Brehe M (2007) Ein Nachfragesystem für dynamische Mikrosimulationsmodelle. In: Dissertation. University of Potsdam

Brunner TA, Siegrist M (2011) A consumer-oriented segmentation study in the Swiss wine market. Brit Food J 113 (3):353–373

Cembalo L, Caracciolo F, Pomarici E (2014) Drinking cheaply: The demand for basic wine in Italy. Aust J Agri Res Econ, doi:10.1111/1467-8489.12059

Christensen LR, Jorgenson DW, Lau LJ (1975) Transcendental logarithmic utility functions. Am Econ Rev 65(3):367–383

Chung C, Dong D, Schmit TM, Kaiser HM, Gould BW (2005) Estimation of price elasticities from cross-sectional data. Agribusiness 21(4):565–584

Corrao G, Rubbiati L, Bagnardi V, Zambon A, Poikolainen K (2000) Alcohol and coronary heart disease: A meta-analysis. Addiction 95(10):1505–1523

Cox TL, Wohlgenant MK (1986) Prices and quality effects in cross-sectional demand analysis. Am J Agr Econ 68(4):908–919

Cranfield JAL, Eales JS, Hertel TW, Preckel PV (2003) Model selection when estimating and predicting consumer demands using international, cross section data. Empir Econ 28(2):353–364

De Boer P, Paap R (2009) Testing non-nested demand relations: Linear expenditure system versus indirect addilog. Stat Neerl 63(3):368–384

Deaton A (1990) Price elasticities from survey data: Extensions and Indonesian results. J Econometrics 44(3):281–309

Deaton A, Muellbauer J (1980a) An Almost Ideal Demand System. Am Econ Rev 70(3):312–326

Deaton A, Muellbauer J (1980b) Economics and consumer behavior. Cambridge University Press, Cambridge

Dey MM, Alam MF, Paraguas FJ (2011) A multistage budgeting approach to the analysis of demand for fish: An application to inland areas of Bangladesh. Mar Res Econ 26(1):35–58

Dey M, Gmel G, Studer J, Dermota P, Mohler-Kuo M (2013) Beverage preferences and associated drinking patterns, consequences and other substance use behaviours. Eur J Public Health, doi:10.1093/eurpub/ckt109

Doran CM, Byrnes JM, Cobiac LJ, Vandenberg B, Vos T (2013) Estimated impacts of alternative Australian alcohol taxation structure on consumption, public health and government revenues. Med J Aust 199(9):619–622

Ecker O, Qaim M (2010) Analyzing nutritional impacts of policies: An empirical study for Malawi. World Dev 39 (3):412–428

Edgerton DL (1997) Weak separability and the estimation of elasticities in multistage demand systems. Am J Agr Econ 79(1):62–79

Estruch R, Ros E, Salas-Salvadó J, Covas M-I, Corella D, Arós F, Gómez-Gracia E, Ruiz-Gutiérrez V, Fiol M, Lapetra J, Lamuela-Raventos Serra-Majem L, Pintó X, Basora J, Muñoz MA, Sorlí JV, Martínez JA, Martínez-González MA (2014) Primary prevention of cardiovascular disease with a Mediterranean diet. New England J Med, doi:10.1056/NEJMx140004

Eurostat (2013) Statistiken zur Nahrungsmittelkette – Vom Erzeuger zum Verbraucher. Press release, EU Commission, Brussels

Fashogbon AE, Oni OA (2013) Heterogeneity in rural household food demand and its determinants in Ondo State, Nigeria: An application of quadratic almost ideal demand system. J Agric Sci 5(2):169–177

FDF (2009) Alkohol in der Schweiz – Übersicht. Swiss Federal Department of Finance, Bern

Fergusson SM, McLeod GG, Horwood JL (2013) Alcohol misuse and psychosocial outcomes in young adulthood: Results from a longitudinal birth cohort studied to age 30. Drug Alcohol Depen 133(2):513–519

Fogarty J (2008) The demand for beer, wine and spirits: Insights from a meat analysis approach., AAWE Working Paper No. 31, American Association of Wine Economists

FOPH (2013) Faktenblatt: Entwicklung des Alkoholkonsum der Schweiz seit den 1880er Jahren. Swiss Federal Office of Public Health, Bern

FOPH (2014a) Alcohol policy regulation in Europe. Swiss Federal Office of Public Health, Bern, http://www.bag.admin. ch/themen/drogen/00039/10172/12019/index.html?lang=en. Accessed 3 Aug 2014

FOPH (2014b) Swiss food composition database. Swiss Federal Office of Public Health, Bern, http://naehrwertdaten.ch/ request?xml=MessageData&xml=MetaData&xsl=Start&lan=en&pageKey=Start. Accessed 3 Aug 2014

Gallet CA (2007) The demand for alcohol: A meta-analysis of elasticities. Aust J Agr Resour Ec 51(2):121–135

Gallet CA (2010) The income elasticity of meat: a meta-analysis. Aust J Agr Resour Ec 54(4):477–490

Gruenewald PJ, Ponicki WR, Holder HD, Romelsjö A (2006) Alcohol prices, beverage quality, and the demand for alcohol: Quality substitutions and price elasticities. Alcohol Clin Exp Res 30(1):96–105

Hagenaars AJM, De Vos K, Zaidi MA (1994) Poverty statistics in the late 1980s: Research based on micro-data. Office for Official Publications of the European Community, Luxembourg

Heckman J (1979) Sample selection bias as a specification error. Econometrica 47(1):153–161

Heeb JL, Gmel G, Zurbrügg CH, Kuo M, Rehm J (2003) Changes in alcohol consumption following a reduction in the price of spirits: a natural experiment in Switzerland. Addiction 98(10):1433–1446

Heien D, Wessels CR (1990) Demand systems estimation with microdata: A censored regression approach. J Bus Econ Stat 8(3):365–371

Holt MT, Goodwin BK (2009) The almost ideal and translog demand systems., MPRA Paper No. 15092, Munich

Jacobs L, Steyn N (2013) Commentary: If you drink alcohol, drink sensibly: Is this guideline still appropriate? Ethn Dis 23(1):110–115

Jithitikulchai T (2011) U.S. alcohol consumption: Tax instrumental variables in Quadratic Almost Ideal Demand System., Paper presented at the Agricultural & Applied Economists Association and Northeastern Agricultural and Resource Economics Association Joint Annual Meeting, Pittsburgh, Pennsylvania, 24–26 July 2011

Klonaris S, Hallam D (2003) Conditional and unconditional food demand elasticities in a dynamic multistage demand system. Appl Econ 35(5):503–514

Kuo M, Heeb JL, Gmel G, Rehm J (2003) Does price matter? The effect of decreased price on spirits consumption in Switzerland. Alcohol Clin Exp Res 27(4):720–725

Lazaridis P (2003) Household meat demand in Greece: A demand systems approach using microdata. Agribusiness 19(1):43–59

Lewbel A, Pendakur K (2009) Tricks with Hicks: The EASI demand system. Am Econ Rev 99(3):827–863

Majumder A, Ray R, Sinha K (2012) Calculating rural–urban food price differentials from unit values in household expenditure surveys: A comparison with existing methods and a new procedure. Am J Agr Econ 94(5):1218–1235

Manning WG, Blumberg L, Moulton LH (1995) The demand for alcohol: The differential response to price. J Health Econ 14(2):123–148

McKelvey C (2011) Price, unit value, and quality demanded. J Dev Econ 95(2):157–169

Mhurchu CN, Eyles H, Schilling C, Yang Q, Kaye-Blakem W, Genç M, Blakely T (2013) Food prices and consumer demand: Differences across income levels and ethnic groups. PLoS One, doi: 10.1371/journal.pone.0075934

Moschini G, Moro D, Green RD (1994) Maintaining and testing separability in demand systems. Am J Agr Econ 76(1):61–73

Nelson JP (1997) Economic and demographic factors in US alcohol demand: A growth-accounting analysis. Empir Econ 22(1):83–102

Oehlert GW (1992) A note on the delta method. Am Stat 46(1):27–29

Perali F, Chavas JP (2000) Estimation of censored demand equations from large cross-section data. Am J Agr Econ 82(4):1022–1037

Pollak RA, Wales TJ (1978) Estimation of complete demand systems from household budget data: The linear and quadratic expenditure systems. Am Econ Rev 68(3):348–359

Reavley NJ, Jorm AF, McCann TV, Lubman DI (2011) Alcohol consumption in tertiary education students. BMC Public Health, doi: 10.1186/1471-2458-11-545

Rehn N, Room R, Edwards G (2001) Alcohol in the European region: Consumption, harm and policies, Report. WHO Regional Office for Europe, Copenhagen

Rice N, Carr-Hill R, Dixon P, Sutton M (1998) The influence of households on drinking behavior: A multilevel analysis. Soc Sci Med 46(8):971–979

SAB (2013) Zahlen und Fakten – Konsum. Swiss Alcohol Board, Bern, http://www.eav.admin.ch/dokumentation/00439/00564/index.html?lang=de. Accessed 3 Aug 2014

Sabia S, Elbaz A, Britton A, Bell S, Dugravot A, Shipley M, Kivimaki M, Sing-Manoux A (2014) Alcohol consumption and cognitive decline in early old age. Neurology, doi: 10.1212/WNL.0000000000000063

Sacks JJ, Roeber J, Bouchery EE, Gonzales K, Chaloupka FJ, Brewer RD (2013) State costs of excessive alcohol consumption, 2006. Am J Prev Med 45(4):474–485

Schwartz LM, Persson EC, Weinstein SJ, Graubard BI, Freedman ND, Mannisto S, Albanes D, McGlynn KA (2013) Alcohol consumption, one-carbon metabolites, liver cancer and liver disease mortality. PLoS One 8(10):e78156

Selvanathan S, Selvanathan EA (2005) The demand for alcohol, tobacco and marijuana: International evidence. Ashgate Publishing, Aldershot

SFSO (2011) Haushaltsbudgeterhebung 2005. Swiss Federal Statistical Office, Neuchâtel

Shonkwiler JS, Yen ST (1999) Two-step estimation of a censored system of equations. Am J Agr Econ 81(4):972–982

Stasi A, Seccia A, Nardone G (2010) Market power and price competition in the Italian wine market. In: Enometrica, Review of the Vineyard Data Quantification Society (VDQS) and the European Association of Wine Economists (FuAWE) - Macerata University

Stasi A, Seccia A, Nardone G (2011) Italian wine consumers' preferences and impact of taxation on wines of different quality and source., Paper presented at the 6th Academy of Wine Business International Conference, France

Su SJ, Yen S (2000) A censored system of cigarette and alcohol consumption. J Appl Econ 32(6):729–737

Tafere K, Taffesse AS, Tamru S (2010) Food demand elasticities in Ethiopia: Estimates using household income consumption expenditure (HICE) survey data. In: ESSP II Working Paper 11. International Food Policy Research Institute, Washington

Tauchmann H (2005) Efficiency of two-step estimators for censored systems of equations: Shonkwiler and Yen reconsidered. J Appl Econ 37(4):367–374

Theil H (1965) The Information approach to demand analysis. Econometrica 33(1):67–87

Thiele S (2008) Elastizitäten der Nachfrage privater Haushalte nach Nahrungsmitteln – Schätzung eines AIDS auf Basis der Einkommens- und Verbrauchsstichprobe 2003. Agrarwirtschaft 57(5):258–268

Thiele S (2010) Erhöhung der Mehrwertsteuer für Lebensmittel: Budget- und Wohlfahrtseffekte für Konsumenten. Jahrb Natl Stat 230(1):115–130

Tobin J (1958) Estimation of relationships for limited dependent variables. Econometrica 26(1):24–36

Van Oers JAM, Bongers IMB, Van de Goor LAM, Garretsen HFL (1999) Alcohol consumption, alcohol-related problems, problem drinking, and socioeconomic status. Alcohol Alcoholism 34(1):78–88

Vermeulen F (2001) A note of Heckman-type corrections in models for zero expenditures. Appl Econ 33(9):1089–1092

Wakabayashi I (2013) Relationship between alcohol intake and lipid accumulation product in middle-aged men. Alcohol Alcoholism 48(5):535–542

WHO (2007) Prevention of cardiovascular disease. Report. World Health Organization, Geneva

Xuan ZM, Nelson TF, Heeren T, Blanchette J, Nelson DE, Gruenewald P, Naimi TS (2013) Tax policy, adult binge drinking, and youth alcohol consumption in the United States. Alcohol Clin Exp Res 37(10):1713–1719

Yen ST, Kan K, Su SJ (2002) Household demand for fats and oils: Two-step estimation of a censored demand system. Appl Econ 34(14):1799–1806

Zheng Z, Henneberry SR (2010) An analysis of food grain consumption in urban Jiangsu province of China. J Agri Appl Econ 42(2):337–355

Zheng Z, Henneberry SR (2011) Household food demand by income category: Evidence from household survey data in an urban Chinese province. Agribusiness 27(1):99–113

8

Country of origin and willingness to pay for pistachios: a chinese case

Pei Xu[1][*] and Zhigang Wang[2]

* Correspondence:
pxu@csufresno.edu
[1]California State University at Fresno,
Department of Agricultural Business,
5245 N Backer Avenue, M/S PB101,
Fresno, CA 93740-8001, USA
Full list of author information is
available at the end of the article

Abstract

Using 360 questionnaire data collected in Beijing, China, this study examines consumers' acceptance and willingness to pay (WTP) for pistachios produced in China, California and Turkey. The impact of country of origin (COO), price, flavor, package size and package type was analyzed. A conditional logit model shows that consumers are willing to pay a statistically significant premium for California and Chinese pistachio, but the Turkish pistachio is less preferred. Using a mixed logit model to estimate the income effect, this study shows that wealthier consumers tend to purchase California pistachio. The marginal effect shows that every 1% increase in income will result in a 1.2% increase in the probability of purchasing California pistachio. However, the purchase of Chinese pistachio does not depend on income.

Price and COO are the two most influential attributes and package size and type are the two least important attributes to change derived utility.

Keywords: Pistachios; Price acceptance; Willingness to pay; Country of origin impact; Conjoint analysis; Pistachio consumption demand in China

Background

Food consumption patterns in China have evolved in the past decade due to expanded consumer disposable income, new desires for Western lifestyle, and heightened food safety concerns (Wang et al. 2008; Xu et al. 2010; Ortega et al., 2011; Xu et al. 2012). With a greater income to spend, the demand for healthful foods with added nutrition features has gone up progressively (Fuller et al. 2006; Wang et al. 2008). The demand for pistachios as a healthful snack food has arisen: as the world's sixth largest grower, China consumes most of its domestically produced pistachios. However, the market share for domestically grown nuts is declining to lower than 50% in 2014, primarily due to food safety concerns (China Commerce, 2013). It was publicized in 2013 that Chinese suppliers used sulphur, a toxic chemical, to color pistachios to make the nuts look appealing. California and Turkey are China's primary international suppliers who ship a total of 12 million pounds of pistachios to China each year (Tech-food.com 2006; PistachioHealth.com, 2012; Askci.com, 2012). China is the outmost export market for these two countries: California sells 17% ($109 million value; a market share of 13%, Enorth, 2008) (South China Morning Post, 2013; Weston, 2013) and Turkey ships 18% of its total crops to China (USDA, 2003). However, anecdotal evidences revealed that California pistachios are unsafe due to salmonella contaminations (Sohu, 2009). To develop efficient marketing strategies and rebuild consumption confidence, international marketers and stakeholders, including those from California, have sought academic help to better understand Chinese consumers' pistachio purchase preference. To compete with foreign suppliers, Chinese domestic pistachio marketers have also looked for academic help to learn factors affecting pistachio purchase choices.

A paucity of published information shows that pistachios have become many city duelers' most favored snack food for which they are willing to spend $10-20 per month, and the quantity demanded is rising (China Food News, 2012). In contrast, a decade ago, pistachios were often presented during major Chinese holidays, or used as a luxury gift to please relatives and close friends for the wish of happiness and good fortune. With Pistachio's health attributes being better understood, it wins the Chinese name: the happy nuts. The growing demand may implicate the need for new market interventions to ensure the supply of safe and desirable pistachios. However, little, if any academic research has dedicated to understand the preferred pistachio features. This study is the first to quantify the impact of price, country-of-origin (COO), product flavor, package-size and package type on pistachio choices.

Different from necessity staples, pistachios are an expensive luxury snack food that offers enjoyment and comfort, and the demand can greatly change with income. Before 1995, when consumer disposable income was small, the demand was found low. Lately, China's rapid economic growth and its strengthened integration to the global economy have propelled a greater need for this luxury snack food. However, none of published work can explain factors driving the demand change and the magnitude of the change. Published food demand analyses in China have exclusively emphasized the demand for necessity staples, such as rice, pork, fish, and milk (Zhang et al. 2010; Ortega et al. 2011; De Steur et al. 2012; Zheng et al. 2012; Xu et al. 2012). These analyses show that Chinese consumers tend to use price as a quality cue and they appear to prefer a higher priced item, assuming it has a better quality. Is price a concern when consume pistachios? A preliminary market research we conducted shows that imported pistachios are priced at 60 Yuan to 180 Yuan per 500 grams ($9.5-28.6, 1$ = 6.3Yuan). The preliminary research did not show that California pistachios are priced higher than domestic produced or imported products from other countries. The package size varies from small size of about 125 grams to large size of about 500 grams. This study analyzes the impact of price on pistachio choice. Findings may shed light to new consumption trend about healthful snack food in general.

This study also differs from published work to take into account the impact of COO on pistachio choice. The literature highlights the impact of consumption experience, retail formats, brands, and government certifications on Chinese consumers' food decisions (Wang et al. 2008; Zhang et al. 2010; Ortega et al. 2011; Xu et al. 2012). Given the substantial presence of imported pistachios from Turkey and from California, including COO as an explanatory variable is necessary. This study analyzes consumer rankings about thirty-six different pistachio profiles to understand: 1) how price and COO affect buying decisions; 2) whether original non-flavored pistachios is preferred to the flavored alternative; 3) how non-gift packed products is liked compared to the gift-packed alternative; and 4) how consumers' income level affects their price, COO, flavor and package preference.

Methods

A summary of selected attributes appears in Table 1. Marketing researchers view COO as a salient cue that affects buyer perceptions of quality (White and Cundiff, 1978). It

Table 1 Selected attributes

Attributes	Levels
Country of origin	China
	California
	Turkey
	Opt-out
Price	80 Yuan/500 grams
	100 Yuan/500 grams
	120 Yuan/500 grams
Flavor	Original flavor (no salt no sweet)
	Sweet flavor
	Salt flavor
Package size	125 grams
	250 grams
	500 grams
Gift pack	Gift pack
	Non-gift pack

also affects product evaluation process and it changes belief formation about a product that has a significant import presence (Erickson et al. 1984). COO was also found to be a deterministic factor affecting food choices and the literature emphasizes beef choices. Schnettler et al. (2004) (cited in Schnettler et al. 2008) showed that imported products were perceived to have lower quality than local beef. Loureiro and Umberger (2007) interviewed 5000 U.S. consumers to examine factors affecting the demand of beef rib-eye steaks. They discovered that COO is less important than a USDA food safety label on beef choices. The COO attribute is critical to pistachio choices and this study addresses the impact of significant presence of imported pistachios.

It is important to help marketers understand whether Chinese consumers prefer sweetened or salted pistachio and how much price premium they are willing to pay to get the preferred taste. Daillant-Spinnler et al. (1996) examined how flavor changes consumption choices of apples and concluded that some consumers preferred sweet and hard apples and some preferred acidic and juicy apples. Blackman et al. (2010) examined taste preference for wine products and found that experienced consumers preferred wine with less sweetness and new consumers favored wine with more sweetness. Though information is limited to understand Chinese consumers' preference for sweetness, information about Asian consumers sweetness preference is available. Comparing the sweetness and saltiness likings of Japanese and Australia consumers, Laing et al. (1994) showed that Japanese consumers agreed with Australia consumers about the sweetness strength of biscuits, jams, chocolates, breakfast cereals, and fruit juice, but they showed great difference in sweetness likings. They rated higher sweetness liking to products produced in their own country. Prescott et al. (1993) studied Japanese and Australia consumers' preference for salty crackers and found that Japanese rated the saltiness intensity similarly as Australia consumers. However, to Japanese, the overall product feature rather than the saltiness content affected the likings rating. Taste may greatly affect pistachio choice and this study asks respondents to rate their preference for saltiness and sweetness to find out how Chinese consumers like the flavored pistachios.

Despite the fact that package greatly influences snack food selection, research related to snack food package preference is scant. Mayen et al. (2007) studied the impact of packaging on Indiana consumers' choice of fresh-cut melon products and found that consumers are willing to pay a price premium for a squared packed product than a cup packed one. The impact of package size and gift- or non-gift package is analyzed in this study to evaluate how packaging affects pistachio willingness to pay.

A choice-based conjoint (CBC) analysis is used to elicit consumers' WTP for pistachios.

This study asked each respondent to rank four alternative products: the first three alternatives feature a different country where the product is from, and the last alternative refers to "I do not like any of the product". The small choice set helps reduce respondents' cognitive burden. Tcornhese four alternatives are randomly selected from a computer generated full factorial design. According to Lusk and Norwood (2005), the random selection maintains orthogonality of the picked attributes; it performs well when the impact is analyzed using discrete utility functions; and it enables the estimation of the impact from interacted attributes. Each respondent completed three choice cards. A sample choice card shows in Table 2. Respondents also completed 32 other questions about consumption frequency and quantity, retail outlets visited, factors affecting choices, and demographics.

Graduate agricultural economics students from a Chinese university administered preliminary face-to-face interviews with Beijing consumers in late December 2011 to find out which factors affect pistachio purchase. These interviews helped select COO, price, sweet or salty, gift package or regular package, and package size as the attributes. Graduate students from the same Chinese university administered the face-to-face interviews in Beijing's seven districts in mid January 2012: Haidian (21%), Chaoyang (23%), Mentougou (4%), Shijingshan (4%), Xicheng (16%), Fengtai (17%), and Dongcheng (15%). Randomly selected respondents completed the survey in large supermarkets (70%), such as Wal-Mart, Chaoshifa Superstores, and Carrefour stores; in small- or mid- size supermarkets (21%), such as Chengnanjiayuan and Quchenshi stores; and in subway stations and randomly selected fruits and nuts stands (9%). More than 90% of the interviews were conducted at the dry nuts section in selected stores. Interviews took place a week before China's Spring Festival when many consumers shop for dry nuts as a holiday gift. Gathering data during this holiday season allowed us to recruit more respondents and we were also able to include in our sample respondents shopping for their own and those shopping for a gift. A total of 360 completed questionnaires were analyzed.

Table 2 A sample choice card

Please circle One product	Country of origin	Price (Yuan/500 grams)	Flavor	Package weight (in gram)	Package type
1	China	120	Salt	500	Gift pack
2	California, U.S.	80	Salt	125	Regular pack
3	Turkey	120	Sweet	250	Gift pack
4	I do not like any				

According to Lancaster's random utility theory (Lancaster 1966), the utility of the ith consumer U_i $(i = 1,...,I)$ derived from the jth product alternative (out of a choice set of C) is a function of the selected attributes of alternative j:

$$U_{ij} = \beta x_{ij} + \varepsilon_{ij} \tag{1}$$

Where β is a vector of unknown parameters of interest, x is a vector of attributes for product j selected by consumer i, and ε is a stochastic error term resulted from measurements errors.

According to McFadden (1974), the probability P_{ij} that an individual i will choose alternative j from choice set C equals the probability that the utility associated with choice j is greater than the utility associated with all other k choices in the same choice set. Thus,

$$P_{ij} = P\left(\beta x_{ij} + \varepsilon_{ij} > \beta x_{ik} + \varepsilon_{ik}\right)$$
$$P_{ij} = P\left(\varepsilon_{ij} - \varepsilon_{ik} > \beta x_{ij} - \beta x_{ik}\right), j \neq k \tag{2}$$

Assume the error terms are independent and identically distributed with the Weibull (Gnedenko, extreme value) distribution (McFadden 1974), P_{ij} is:

$$P_{ij} = \frac{exp\left(\beta x_{ij}\right)}{\sum_{k=1}^{j} exp(\beta x_{ik})} \tag{3}$$

In the above conditional logit model, x represents product attributes only. The model assumes the characteristics of respondents are the same across the sample.

In order to control heterogeneity of respondents' individual specific characteristics, a mixed logit model is applied. This model allows the interaction between individual specific characteristics and product attributes in the estimation (Train, 2003; Colombo et al. 2007). For individual i, the random parameters β can be specified as:

$$\beta \sim H(\theta, v) \tag{4}$$

where H(.) is a probability distribution function with mean θ and variance v of the underlying distribution function. The probability of individual i choose alternative j is given by:

$$P_{ij} = \int \frac{exp\left(x_{ij}\beta\right)}{\sum_{k=1}^{j} exp(x_{ik}\beta)} \delta(\beta)d\beta \tag{5}$$

where $\delta(\beta)$ is the joint density function for the random parameter β (Hu et al. 2009; Chang et al. 2012). Willingness to pay (WTP) estimates the amount of money an individual consumer is willing to give up to exchange for benefits of a specific product attribute. According to Lusk and Norwood (2005), Mayen et al. (2007) and Chang et al. (2012), WTP can be computed using:

$$WTP_j = \frac{\beta_{j=1} - \beta_{j=0}}{-\beta_{price}} \tag{6}$$

The variance of WTP is estimated using (Greene, 2000):

$$var[WTP] \approx \left(\frac{\partial WTP}{\partial \beta}\right)' \left(var\left[\hat{\beta}\right]\right) \left(\frac{\partial WTP}{\partial \beta}\right) \qquad (7)$$

The gathered data were analyzed using the following model (McFadden, 1974).

$$\begin{aligned} Utility = \;& \beta_1(California) + \beta_2(China) + \beta_3(Turkey) \\ & + \beta_4(Price) + \beta_5(salt) + \beta_6(sweet) \\ & + \beta_7(gift) + \beta_8(midsize) + \beta_9(largesize) \end{aligned} \qquad (8)$$

The first three coefficients in equation (8) estimate the impact of COO on derived utility. These three coefficients are relative to the coefficient of the opt-out alternative. The impact of the salt and sweet attributes was relative to the original favor alternative, which was the base and was omitted in the estimation. The gift attribute's impact was estimated relative to the non-gift package alternative, which was omitted, too. The medium size (250 grams) and large size (500 grams) attributes' impact was relative to the small size (125 grams) alternative, which was omitted as well. Respondents were expected to select the lower priced alternative rather than the higher priced one, resulting in a negative price coefficient. STATA 11 econometric software was used to estimate the parameters.

Results and discussion

Respondents' demographics appear in Table 3. The sample includes people from different age, gender, education and income groups. A majority of these respondents are 40 years or younger (72%) and more females answered the survey (55%). Given the fact that interviews were mostly conducted at the dry nuts section, it seems to suggest that younger females are more likely to visit this section. The sample's mean monthly income is 6000 Chinese Yuan ($ 952; 1 USD = 6.3 Chinese Yuan) with 55% respondents reported an income of more than 6000 Yuan. The sample's average household size is 3.25 with 65% respondents reported a household size of three or fewer. This household size statistics is higher than the published national average household size statistics of 2.45 (The Six National Population Census, Beijing, 2014). The sample's per capita income of 1846 Yuan ($293) is a little lower than the published Beijing per capita income, which was $317 in 2011 (Beijing Statistical Information Net, 2011). The sample's education level is higher than the 2011 published Beijing average education level which shows that 20.56% of the Beijing population have a bachelor's degree (Chen et al. 2006). Our sample shows that 34% of the respondents have a bachelor's degree. 70% of the respondents reported they do not have a family member younger than 14 and 55% of them reported that they do not have a family member older than 50. Thus, many of the respondents do not have a dependent child and they also do not have elderlies to support.

Table 4 shows the reported pistachio purchase behavior. Fifty four percent of the respondents would purchase 500 grams or more each time they visit a store. Among them, 25% would purchase 1500 grams or more. This high purchase quantity comes with a high purchase frequency: 31% of them make a purchase once every month or more often. Though the sample seems to suggest a high consumption demand, it also reveals that 38% of the respondents make a purchase only when the family celebrates major holidays; and 19% of them do not buy pistachios. However, 31% of the

Table 3 Respondents' demographics

(n = 360)	Frequency	Percentage
Age		
18-25	108	30%
26-40	152	42%
41-55	57	16%
Above 55	26	7%
No response	17	5%
Total	360	100%
Gender		
Male	159	44%
Female	198	55%
No response	3	1%
Total	360	100%
Household Monthly Income (Chinese Yuan)		
2500 or less	35	10%
2501-6000	128	35%
6001-9000	67	19%
Above 9000	130	36%
Total	360	100%
Household member		
Two or fewer	75	21%
Three	159	44%
Four	84	23%
Above four	42	12%
Total	360	100%
Education		
High school or below	82	23%
Job training school	85	23%
Bachelors	122	34%
Graduate	71	20%
Total	360	100%
Household with a member younger than 14		
None	253	70%
One	94	26%
Two	10	3%
Three	3	1%
Total	360	100%
Household with a member older than 50		
None	199	55%
One	40	11%
Two	115	32%
Three	5	2%
No response	1	0
Total	360	100%

Table 4 Pistachio purchase behavior

(n = 360)	Frequency	Percent
Purchase quantity/store visit (grams)		
500 or less	167	46%
501-1000	87	24%
1001-1500	17	5%
Above 1500	89	25%
Total	360	100%
Purchase frequency		
Once per month or more often	110	31%
Once every two to three month	45	12%
When celebrate major holidays only	135	38%
Do not purchase pistachio nuts	70	19%
Total	360	100%
Compare to last year, this year your family purchased more pistachios or less?		
More	65	22% (20% increase)
Not change	196	68%
Less	29	10% (25% decrease)
Total	290	100%
Often purchase domestic produced or imported?		
Domestic produced	117	40%
Imported	40	14%
Both	79	27%
Do not know country of origin	54	19%
Total	290	100%
Often purchase for family use or as a gift to others		
For family	233	80%
Use as a gift	10	3%
Both	45	15%
I do not know	2	1%
Total	290	100%
Where do you often shop for pistachios?		
Large supermarkets	169	58%
Midsize- small- supermarkets	62	21%
Dry nuts wholesale store	30	10%
Dry nuts retail store	18	6%
Wet markets	7	2%
Online store	4	1%
Total	290	100%
Have you heard about California pistachios		
Yes	226	74%
No	81	26%
Total	307	100%
Have you purchased California pistachios		
Yes	150	52%
No	138	48%

Table 4 Pistachio purchase behavior *(Continued)*

Total	288	100%
Do you like to try California pistachios		
Yes	154	94%
No	10	6%
Total	164	100%

respondents do make a purchase at least once per months and the average consumption amount for this group is 640 gram per visit. This group consumes pistachio on a daily basis. Our sample also suggests that compared to the last year (year 2010), 22% of the respondents have increased the purchase by 20%; 68% have purchased a similar amount; and 10% have decreased the purchase by 25%. Though the interview did not ask reasons behind the consumption change, possible explanations include change in income and food preference, food safety concerns, and newly availability of imported pistachios.

COO appears to affect pistachio purchase: 40% of the respondents often purchase domestically produced pistachio only; 14% prefer imported nuts only; 27% purchase both; and 19% do not know the COO of the pistachios. Thus, the statistics seem to indicate that domestically produced nuts are more popular than imported nuts. Most purchase was made for own consumption only (80%). Only a few purchases were made for gift purpose (3%) and more purchase was made for both occasions (15%). Thus, pistachios are primary used for own consumption. This finding is out of our expectation as we expect more purchase would be made for gift purposes, especially during the holiday seasons. The result possibly indicates that pistachio is not any more a luxury item to entertain friends or relatives during major holidays. Rather it is considered as a daily snack to be consumed by respondents' own family.

Most purchases took place at supermarkets of different size (79%) with only 16% purchases were made at wholesale or retail dry nuts store. Purchase is less likely to be made at wet markets or online stores (3%). Many respondents have heard about California produced pistachio (74%); 52% of them have purchased California pistachio; and 94% would like to try California pistachios in the future.

Results of the main effect model appear in Table 5. The likelihood ratio test, the Prob > Chi^2 score, and the Wald chi^2 results suggest that the conditional logit model is a good fit. Specifically, the likelihood ratio test rejects the null hypothesis that none of the selected attributes contribute to the derived utility. At least one of the selected attributes has a statistically significant impact on the estimated utility. Coefficient estimates of the conditional logit model describe the effect of each selected attribute on the overall derived utility of pistachio consumption. Coefficients for the California alternative and the Chinese alternative are both positive and statistically significant (alpha < 0.0001), meaning that pistachio from these two countries are greatly preferred over the no purchasing option. Thus, the sample suggests that responding Chinese consumers appreciate Chinese pistachio as well as imported pistachio from California. However, the bigger coefficient of the California variable than the Chinese variable further suggests that California pistachio adds more to the derived utility than the Chinese alternative. A likelihood ratio test was then conducted to find out whether Chinese consumers like California pistachio better than Chinese pistachio in a statistically significant way.

Table 5 Conditional logit model coefficients

Variables	Coefficient	P > |z|
China***	0.7311	<0.0001
California***	1.077	<0.009
Turkey***	−0.4739	<0.0001
Price***	−0.1913	<0.0001
Salty Flavor	−0.0135	0.895
Sweet Flavor***	−0.3856	<0.0001
Gift Pack**	−0.1677	0.049
250-gram Pack	0.0192	0.854
500-gram Pack	−0.1208	0.252
Likelihood ratio	−1361	
Wald chi2(6)	35.34	
Prob > chi2	<0.0001	

Standard errors are in parentheses; *** means statistically significant at 1% level; ** means 5% level.

The LR statistics is 187.6 with 1 degree of freedom, indicating that Chinese consumers prefer California pistachio significantly better. The statistically significant negative coefficient of the Turkey variable suggests that the no-purchase option is preferred to the purchase of Turkish pistachio. Thus, California pistachio is most preferred, followed by the Chinese alternative, and the Turkish pistachio is least favored.

The negative and statistically significant price coefficient indicates rational consumption behavior: Chinese consumers prefer the lower priced items rather than the high priced alternatives. In contrast, when researching food consumption in China, several researchers have found that Chinese consumers' purchasing behavior contradicts with the demand theory: the higher priced items often result in greater quantity demanded, due to heightened food safety concerns (Herrmann et al. 2006; Wang 2003; Xu et al. 2010). For example, Xu et al. 2010 showed that Chinese college students prefer higher priced milk because they believe the milk is safer than lower priced options. Different from milk, which is consumed on a daily basis, to most respondents, pistachio is less frequently consumed. In addition, milk contaminations are publicized so frequently such that it has been difficult to restore consumption confident (Xu et al. 2010). In contrast, pistachio contamination was once uncovered in 2013, but later on more news have emphasized the healthful and beneficial features of pistachios. Thus, food safety appears less a concern and price is still a primary factor changing pistachio choices.

The sweet and salty flavor was compared with the original non-flavor to understand if Chinese consumers prefer original pistachio to the flavored alternatives. The original flavor is significantly preferred to the sweet flavor, as suggested by the negative coefficient of the Sweet Flavor variable. Though the salty flavor is not significantly preferred to the original flavor, both salty- and original- flavored are more attractive than the sweet flavored pistachios.

Our sample also suggests that a non-gift packed product is preferred to the gift-packed alternative. This finding further confirms the fact that pistachio is popularly used for own-consumption instead of a gift. Consumers desire the non-gift packed rather than the more expensive gift-packed alternatives. This finding again indicates how price sensitive Chinese consumers are when selecting pistachios. However, Chinese consumers are found to be less sensitive to package size change. The two package-size variables are not significant, indicating that consumers like the 125-gram pack the same as they like the 250-gram or the 500-gram pack. Small packs are convenient which may especially attract younger consumers. More than ten years ago, Veeck and Veeck (2000) discovered that younger consumers with higher disposable income but little shopping time would value the convenience of food. The small pack may perfectly meet the needs of this group. The larger 500-gram pack, in contrast, may fit the needs of older consumers, who have a bigger family to support and who are more sensitive to price. Given larger packs usually come with a quantity discount, this older consumer group tends to favor the larger packs.

Following Mayen et al. (2007), the relative importance of each selected attribute over total importance were calculated (Table 6). The observed R.I. and the Bootstrap standard errors are computed using 500 random samples generated for the simple regression model (Guan, 2003). Price is the outmost important attribute that contributes to 74% of the total importance. This result again confirms the great impact of price on pistachio purchase. COO adds 17% to the total importance, indicating that consumers look at COO when making pistachio purchases. Flavor, package size, and package type are less important features. Thus, price and COO are two key attributes to change pistachio choice.

Marginal effects estimate how the change of attribute level affects the probability of a product being selected (Table 7). The marginal effect of price is –4.55 (alpha < 1%), indicating that a one dollar increase in price will result in a 4.55% decrease in the probability of selecting California pistachio. The price impact is smaller for domestic Chinese pistachios (3.82%) and for Turkey pistachios (1.45%). Thus, compared to California and Chinese pistachios, the demand for Turkey pistachios is the least sensitive to price change. However, the demand for Turkey pistachio is also the smallest. The marginal effect for the Opt-out alternative is 3.59%, indicating that if price rises, it will make it even more unlikely for the non-purchasers to buy pistachios.

Pistachio flavor and package type affect purchase probability. A move from the original to sweet flavor reduces the probability of choosing California pistachio by 9.01% and Chinese pistachio by 7.47% (alpha < 1%). A change from gift pack to non-gift pack increases the probability of selecting California pistachios by 3.98% (alpha < 5%), and Chinese pistachios by 3.35% (alpha < 5%). The salty flavor and the package size attributes are not resulted in statistically significant marginal effects.

The WTP estimates are shown in Table 8. Surveyed Chinese consumers are willing to pay an additional 6.30 Yuan (equivalent $1, exchange rate: $1 = 6.3 Yuan) on average to move from Turkey to Chinese pistachios, and 8.11 Yuan ($1.3) to move from Turkey

Table 6 Relative importance of selected attributes

	R.I.	Observed R.I.	Bootstrap standard error
Price	0.74	0.92	0.035
COO	0.19	0.17	0.046
Flavor	0.04	0.045	0.02
Gift Pack	0.02	0.02	0.013
Pack size	0.01	0.02	0.015

Table 7 Marginal effects for conditional logit model

Attributes	California	China	Turkey	Opt-out
Price	−4.55 (1.25)***	−3.82(1.06)***	−1.45(0.42)***	−3.59(0.99)***
Origin to Salty	−0.32(2.40)	−0.27 (2.03)	−0.10(0.77)	−0.25(1.91)
Origin to Sweet	−9.01 (2.42)***	−7.47(1.99)***	−2.78(0.76)***	−6.53(1.62)***
Gift to None-gift	3.98(2.02)**	3.35(1.71)**	1.27(0.65)*	3.01(1.47)**
125- to 250-gram	0.45(2.47)	0.38(2.08)	0.14(0.79)	0.36(1.97)
125- to 500-gram	−2.86(2.48)	−2.3(2.07)	−0.90(0.77)	−2.19(1.86)

Standard errors are in parentheses; *** means statistically significant at 1% level; ** means 5% level; * means 10% level.

to California pistachios. The 95% confidence interval does not include a zero, indicating that the WTP estimates are significantly different from zero. Thus, respondents' WTP is the highest for California pistachios, followed by the Chinese option and the Turkish pistachios the lowest. Surveyed Chinese consumers are willing to pay a higher premium for their preferred California or Chinese pistachio than the Turkey option.

Chinese consumers are willing to pay a small premium for pistachio with original, no-salt and no-sugar flavor. The average WTP is 2.02 Chinese Yuan ($ 0.3) to move from sweet pistachios to an original product. The WTP estimate is not statistically significant for the Original to Salty variable, meaning Chinese consumers do not prefer original flavor to salty flavor. Combining this result with results in Table 7, we conclude that original or salty flavor is more attractive than the sweet flavor. WTP estimates are not significant for the package type and package size variables, indicating that respondents' WTP will not change with package type and package size. The WTP estimates are the greatest for the COO attributes, followed by the original flavor attributes. In addition, respondents will not pay a premium for preferred package type or package size. This result agrees with the relative importance estimates in Table 6. The WTP estimate for package type in Table 8 is insignificant, but the coefficient estimate for package type is significant in Table 5. This seems to suggest that respondents consider package type when purchasing pistachio, however, they will not pay a premium for pistachios that have a preferred package type.

Consumer characteristics were found to interact with product attributes to change choice preference (Louviere et al., 2000). This study further addresses the interaction effects and uses a mixed logit model to estimate how COO effect would change with income (Table 9). The likelihood ratio test, the Prob > chi^2 score, and the Wald chi^2 results all suggest that the mixed logit model is a good fit. COO, price, sweet flavor and pack type that were significant in the conditional logit model remain significant in the

Table 8 WTP estimates for conditional logit model

Change in attributes	Mean WTP	Standard deviation	95% confidence interval	
Turkey to Chinese	6.3	1.85	2.67	9.93
Turkey to California	8.11	2.3	3.6	12.61
Original to salty	0.07	0.53	−0.98	1.12
Sweet to original	2.02	0.78	0.48	3.55
Gift to non-gift pack	0.88	0.51	−0.12	1.88
125- to 250-gram	0.1	0.54	−1.17	0.97
125- to 500-gram	−0.63	0.58	−0.5	1.76

Table 9 Mixed logit model coefficients

Variables	Coefficient	P > |z|
China**	0.5261	0.049
California**	0.5132	0.044
Turkey***	−0.4689	0.01
Price***	−0.1929	<0.0001
Salt Flavor	−0.0034	0.973
Sweet Flavor***	−0.3947	<0.0001
Gift Package**	−0.1716	0.045
250-gram Pack	0.0212	0.839
500-gram Pack	−0.1272	0.229
Chinese X Income	0.0192	0.332
California X Income***	0.0513	0.005
Likelihood ratio	−1353	
Wald chi2 (8)	44.02	
Prob > chi2	<0.0001	

Standard errors are in parentheses; *** means statistically significant at 1% level; ** means 5% level.

mixed logit model. The California and income interaction effect is positive and statistically significant at 1% level, suggesting that the purchase of California pistachios depends on income. Respondents with a higher income tend to purchase California pistachios. The marginal effect shows that: every 1% increase in income will result in a 1.2% increase in the probability of choosing California nuts. The Chinese and income interaction effect is positive but not statistically significant, meaning that the higher income will not generate the purchase of Chinese pistachios. The COO impact on derived utility is reduced in this income effect model compared to the main effect model (0.7311 to 0.5261 for China and 1.077 to 0.5132 for California). Thus, wealthier respondents tend to derive additional utility from the consumption of California pistachios.

WTP for the mixed logit model appears in Table 10. On average, surveyed consumers are willing to pay an extra of 5.16 Yuan ($0.82) for Chinese pistachios, and 5.09 Yuan ($0.81) more for California pistachios compared to the Turkish alternative. WTP is the highest for Chinese pistachios, followed by the California alternative, and the Turkish pistachios resulted in the lowest WTP. WTP estimates for the Sweet to Original variable is significantly different from zero, meaning that consumers value original flavored

Table 10 WTP estimates for mixed logit model

Change in attributes	Mean WTP	Standard deviation	95% confidence interval	
Turkey to Chinese	5.16	1.9	1.44	8.88
Turkey to California	5.09	1.84	1.49	8.69
Original to Salty	0.02	0.53	−1.02	1.06
Sweet to Original	2.05	0.79	0.51	3.59
Gift pack to None-gift	0.89	0.51	−0.11	1.89
125- to 250- gram pack	0.11	0.54	−1.17	0.95
125- to 500- gram pack	−0.66	0.58	−0.47	1.79
Chinese * Income	0.1	0.1	−0.1	0.3
California * Income	0.27	0.09	0.08	0.45

pistachios. The income interaction indicates that wealthier respondents are willing to pay an additional 27 cents for California pistachios. WTP estimates for the flavor variable, the package type and package size variables are not significant, meaning that respondents' purchase choice is less likely to be affected by these attributes.

Conclusions

China's already expanded pistachio demand will continue to grow due to its strong economic development and the presence of wealthy consumer segments. The greater disposable income enables these new consumption segments to purchase imported California pistachios. These emerging affluent consumers are found to be willing to pay a price premium for California pistachios. Moreover, their demand is found to be rational, meaning that they are willing to pay a higher price for California nuts than Chinese nuts, but they are only interested in reasonably priced California nuts. Thus, California pistachios are attractive only if their price is comparable to that of domestic Chinese nuts. In terms of preferred flavor, the affluent segments want the no-salt and no-sugar flavor or the salty flavor more than the sweetened nuts. To them, non-gift pack seems to be more attractive than the gift pack, indicating that pistachios are popularly consumed in the family rather than exclusively used as a gift.

China's pistachio consumers are increasingly diversified to include the less wealthy consumers. Similarly to the rich group, this less affluent segment cares product price, values product origin, and wants the original flavored pistachios. Perhaps due to limited income this segment prefers domestic Chinese pistachios. The diversified consumption group may predict a demand bust and a great market opportunity to domestic as well as international pistachio shareholders. Our results suggest that California pistachio suppliers should take advantage of the high price willingness to pay from the wealthier consumers. Domestic Chinese suppliers may focus on the less affluent groups to address their needs of lower priced pistachios.

COO as a salient quality cue is found to affected Chinese consumers' pistachio needs. Chinese consumers favor California pistachios the most and Turkish pistachios the least. Domestic Chinese pistachios are more desirable than Turkish nuts, perhaps due to better product knowledge, positive consumption experiences, and easier market accesses. In the Western behavioral literature, better product knowledge explains enhanced preference rating (Bilkey and Nes, 1982). In the Chinese setting, American Pistachio Growers Association showcased the healthfulness of California pistachios in China's two largest pistachio consumption cities of Beijing and Shanghai in 2011 (Cq people.cn, 2011). This and other market promotion programs may have acquainted Chinese consumers with California pistachios and helped California pistachios establish a greater market share.

Competing interests
Both authors declare that they have no competing interests.

Authors' contributions
PX and ZW designed the Questionaire; ZW gathered data; PX analyzed data and drafted the manuscript. PX and ZW revised the manuscript. Both authors read and approved the final manuscript.

Author details
[1]California State University at Fresno, Department of Agricultural Business, 5245 N Backer Avenue, M/S PB101, Fresno, CA 93740-8001, USA. [2]School of Agricultural Economics and Rural Development, Renmin University of China, Haidian, Beijing 100091, People's Republic of China.

References

Askci.com (2012) China Imported Olive Oil from Turkey Would Arrive 464 Tons in 2012., http://www.askci.com/news/201206/06/9301_55.shtml Assessed 4 June 2014

Beijing Statistical Information Net (2011)., http://www.bjstats.gov.cn/sjfb/bssj/jdsj/2011/201109/t20110914_210867.htm Assessed 4 June 2014

Bilkey W, Nes E (1982) Country of Origin Effects on Product Evaluations. J Int Bus Stud 8(Spring/Summer):89–99

Blackman J, Saliba A, Schmidtke L (2010) Sweetness Acceptance of Novices, Experience Consumers and Winemakers in Hunter Valley Semillon Wines. Food Qual Prefer 21:679–683

Chang KL, Xu P, Underwood K, Mayen C, Langelett G (2012) Consumers' Willingness to Pay for Locally Produced Ground Beef: A Case Study of the Rural Northern Great Plains. J Int Food Agribus Mark 25(1):42–67

Chen G, Cao GY, Liu YB, Pang LH, Zhang L, Ren Q, Wang HT, Zheng XY (2006) The Future Population of Beijing – A Projection on the Population, Human Capital and Urbanization Using PDE Model. J Mark Popul Anal 12(4):29–41

China Food News (2012) Sales Management and Brand Management to Expand the Pistachio Market., August 2, 2012. http://www.cnfood.cn/npage/shownews.php?id=8838 Assessed 4 June 2014

China Commerce (2013) Food Safety: The China Dream., http://finance.sina.com.cn/leadership/mroll/20130407/103515062173.shtml Accessed 5 June 2014

Colombo SJ, Calatrave-requena J, Hanley N (2007) Testing Choice Experiment for Benefit Transfer with Preference Heterogeneity. Am J Agr Econ 89:135–151

Cq People.cn (2011) Pistachio Grower Association and Lady California Promoted U.S. Pistachio to Beijing Consumers., http://cq.people.com.cn/newscenter/cq/news.cqr300?Num=7467833. Assessed 4 June 2014

Daillant-spinnler B, MacFie HJH, Beyts PK, Hedderley D (1996) Relationships between Perceived Sensory Properties and Major Preference Directions of 12 Varieties of Apply from Southern Hemisphere. Food Qual Prefer 7:112–126

De Steur H, Gellynck X, Feng SY, Rutsaert P, Verbeke W (2012) Determinants of Willingness-to-pay for GM Rice with health Benefits in a High-risk region: Evidence from Experimental Auctions for Folate Biofortified Rice in China. Food Qual Prefer 25:87–94

Erickson GM, Johansson JK, Chao P (1984) Image Variables in Multi-attribute Product Evaluations: Country-of-Origin Effects. J Consum Res 11(2):694–699

Enorth (2008) California Pistachio Seeks to Increase its Market Share in China., http://economy.enorth.com.cn/system/2008/05/29/003328840.shtml. Assessed 4 June 2014

Fuller F, Huang J, Ma H, Rozelle S (2006) Got Milk: The Rapid Rise of China's Dairy Sector and Its Future Prospects. Food Policy 31:201–215

Greene WH (2000) Econometric Analysis, 5th edn. Prentice Hall

Guan WH (2003) From the Help Desk: Bootstrapped Standard Errors. The Stata Journal 3(1):71–80

Herrmann M, Xu P, Dong LC, Fong QS, Crapo C (2006) Rating Alaska Salmon Protein Concentrate in China. J Food Prod Mark 12(1):57–85

Hu W, Woods T, Bastin S (2009) Consumer Acceptance and Willingness to Pay for Blueberry Products with Nonconventional Attributes. J Agr Appl Econ 41(1):47–60

Laing DG, Prescott J, Bell GA, Gillmore R, Allen S, Best DJ, Yoshida M, Yamazaki K, Ishii-Mathews R (1994) Responses of Japanese and Australians to sweetness in the context of different foods. J Sens Stud 9(2):131–155

Lancaster K (1966) A New Approach to Consumer Theory. J Polit Econ 74:132–157

Louviere J, Hensher D, Swait J (2000) Stated Choice Methods: Analysis and Applications, 1st edn. Cambridge University Press, Cambridge, U.K.

Lusk JL, Norwood FB (2005) Effect of Experimental Design on Choice-Based Conjoint Valuation Estimates. Am J Agr Econ 87(3):771–785

Loureiro ML, Umberger WJ (2007) A Choice Experiment Model for Beef: What U.S. Consumer Responses Tell Us About Relative Preferences for Food Safety, Country-of-Origin Labeling and Traceability. Food Policy 32:496–514

Mayen C, Marshall MI, Lusk JL (2007) Fresh-cut Melon-The Money Is in the Juice. J Agr Appl Econ 39(3):597–609

McFadden D (1974) Conditional Logit Analysis of Qualitative Choice Behavior. In: Zarembka P (ed) Frontiers of Econometrics. Academic Press, New York

Ortega DL, Wang HH, Olynk N, Bai J, Wu L (2011) Chinese Consumers Demand for Food Safety Attributes: A Push for Government and Industrial Regulations. Am J Agr Econ 94(2):489–495

PistachioHealth.com, (2012) World Pistachio Day: A Global Chance to Celebrate the Wonderful Green Nuts., http://wwwprnewswirecom/news-releases/world-pistachio-day-a-global-chance-to-celebrate-the-wonderful-green-nut-85334192html Assessed 4 June 2014

Prescott J, Bell GA, Gillmore R, Yoshida M, Laining DG, Allen S, Yamazaki K (1993) Responses of Japanese and Australians to Saltiness in the context of foods. Chem Senses 18(5):616

Schnettler B, Manquilef O, Miranda H (2004) Attributos valorados en la seleccion de carne bovina en supermercados de Temuco, IX Region de Chile. Sci Agri Res 31(2):91–100

Schnettler B, Ruiz D, Sepulveda O, Sepulveda N (2008) Importance of the Country of Origin in Food Consumption in a Developing Country. Food Qual Prefer 19:372–382

Sohu (2009) China's Food Safety Inspection Bureau: China Rejected U.S. Produced Pistachio., http://baobao.sohu.com/20090411/n263326197.shtml Assessed 4 June 2014

South China Morning Post (2013) Chinese Consumers Go Crazy for California Pistachios., http://www.scmp.com/news/hong-kong/article/1295877/chinese-consumers-go-crazy-californian-pistachios Assessed 4 June 2014

Tech-Food.com (2006) China Leads the World's Pistachio Consumption., in Chinese http://wwwtech-foodcom/news/2006-11-21/n0083060htm Accessed 2 March 2013

The Six National Population Census, Beijing (2014)., http://www.bjstats.gov.cn/rkpc_6/pcsj/201105/t20110530_203327. htm Accessed 2 June 2014

Train K (2003) Discrete Choice Methods with Simulation. Cambridge University Press, Cambridge, U.K.

USDA (2003) World Pistachio Situation and Outlook., http://www.agrostrat.gr/?q=en/node/279. Assessed 4 June 2014

Veeck A, Veeck G (2000) Consumer Segmentation and Changing Food Purchase Patterns in Nanjing, PRC. World Dev 28(3):457–471

Wang ZG (2003) Consumers' recognition of food safety and their decision making for consumption. Chinese Rural Eco 4:41–51

Wang Z, Mao Y, Gale F (2008) Chinese Consumer Demand for Food Safety Attributes in Milk Products. Food Policy 33:27–36

Weston S (2013) China Consumes 17% of California's Pistachio Crop., http://www.foodbev.com/news/china-consumes-17-of-californias-pistach#.UnvkKZRVR8s. Assessed 4 June 2014

White P, Cundiff ED (1978) Assessing the Quality of Industrial Products. J Mark 42(January):80–86

Xu P, Zheng S, Motamed M (2010) Perceived Risks and Safety Concerns about Fluid Milk among Chinese College Students. Agri Econ – Czech 56(2):67–78

Xu P, Fong QC, Zeng YC, Lone T, Liu YY (2012) Chinese Consumers' Willingness-to-Pay for Green- and Eco- Labeled Seafood. Food Control 28:74–82

Zhang CP, Bai JF, Lohmar BT, Huang JK (2010) How Do Consumers Determine the Safety of Milk in Beijing, China? China Econ Rev 21:545–554

Zheng S, Xu P, Wang Z, Song S (2012) Willingness to Pay for Traceable Pork in Beijing, China. China Agri Econ Rev 4(2):200–215

9

Case study analysis on supplier commitment to added value agri-food supply chains in New Zealand

Nic J Lees[*] and Peter Nuthall

* Correspondence:
nic.lees@lincoln.ac.nz
Faculty of Agribusiness and
Commerce, Lincoln University,
Ellesmere Junction Road,
Christchurch 7647, New Zealand

Abstract

This research identifies what attracts suppliers to be committed to long-term relationships in New Zealand agri-food supply chains where suppliers are required to consistently deliver to high product specifications. It also looks at what factors determine supplier's ongoing commitment and how to build strong enduring supply chain relationships. Semi structured interviews were undertaken with suppliers from New Zealand agri-food exporting companies. The main factors that attracted suppliers to these supply chains were; increased price certainty, premium prices and relationship quality. Many suppliers wanted to break away from the agricultural commodity cycle, which they saw as disconnected from customer demand, and characterised by price volatility. They saw themselves as better than average producers with the ability to produce high quality products. They valued the relationship with the companies they supplied as this gave them access to premium markets where they felt they would be rewarded for their effort. There was a high level of trust in these relationships and this was built on openness and transparency in communications and confidence in the character of the company personnel. The success of differentiated agri-food supply chains requires capable and committed suppliers. Companies that are developing a differentiated strategy need to identify suppliers who have the ability to produce high quality products and want to be involved in a customer focused supply chain enables them to access to premium markets.

Jel Codes: Q13

Keywords: Supplier relationships; Commitment; Trust; New Zealand; Competitive advantage; Resource based view; Social capital

Background

The New Zealand economy is highly dependent on agri-food exports and is unique among the world's developed economies in that nearly two thirds of exports come from the agricultural sector. For example, Denmark and the Netherlands are the nearest comparable developed economies with significant agri-food export sectors, yet their agri-food exports represent only around 20% of these countries' total exports. The most significant of New Zealand's agri-food exports are dairy and red meat products. The dairy sector generated US$ 10.7 billion in export earnings in 2013, representing 28 per cent of total merchandise export value; while the red meat sector generated US $ 4.2 billion in export earnings (Statistics New Zealand 2013). New Zealand's efficient pasture based production system and small population provide a low cost competitive advantage in the export of high quality meat and dairy products. This dependence, however, makes New Zealand vulnerable to changes in foreign government's policies and consumer demand in the importing countries, as well as competition from other low cost agri-food exporters.

New Zealand has traditionally relied on this low cost competitive advantage (Porter 1998) and focused on improving productivity and efficiency to preserve its position as one of the world's most efficient agricultural producers. This is now becoming more difficult to maintain with rising production costs and regulatory constraints on agricultural intensification. Because of this, many people are questioning if New Zealand still has a sustainable long-term, low cost competitive advantage. The alternative to maintaining this low cost position would be focusing, instead, on increasing the value of the product (Porter 1985b). This would require a fundamental shift in the focus of New Zealand agriculture. Instead of an emphasis on efficient farm production and increasing scale, the focus would need to be on meeting the needs of selected high value consumers. These consumers are demanding greater variety and quality in the food they eat. They require a consistent year-round supply of high quality, safe food (Fischer et al. 2009; Van der Vorst 2000). They also want food that aligns with their own personal values, which includes credence attributes such as environmental sustainability, animal welfare and fair trade, as well as local and organic production.

Meeting these consumer demands is difficult within the constraints of New Zealand's pasture based agricultural production system, where production volume and product specifications are highly dependent on climate. This leads to a fundamental question. Should New Zealand agriculture continue to focus on low cost, efficient production systems? Or, should it focus instead on developing higher value products, with innovative production systems that can deliver a consistent year-round supply of high quality, safe food and also address consumers' concerns for animal welfare and environmental stewardship? This change would be a significant challenge for the relationships in the supply chain. The New Zealand agricultural sector has traditionally relied on short-term spot market exchange relationships (McLeod et al. 2011). While these are efficient for large volumes of undifferentiated products they are less effective in meeting consumer needs for differentiated products (Sonka 2003). In a spot market transaction there is little information flow. Information flow is important with differentiated products where credence quality attributes, such as animal welfare are not visible in the physical product at purchase or, even, after consumption (Nelson 1970; Dyer and Singh 1998). Therefore, to meet these consumer needs the New Zealand agricultural industry would need to move away from relying

Table 1 Proprtion of NZ products exported

Product	Per cent exported	Main market	Per cent to main market 2013
Dairy products	97 per cent	China	32 per cent
Sheep meat	90 per cent	European Union	44 per cent
Beef	80 per cent	USA	43 per cent
Venison	90 per cent	European Union	76 per cent

(Statistics New Zealand 2013).

predominantly on a traditional commodity model with short-term, competitive, spot market relationships to a partnership model with increased supply chain commitment involving long-term contracts and to delivering of high quality products to meet customer demands (Fischer et al. 2009).

This would require suppliers who are willing to commit to meeting higher product specifications while working with less flexible delivery schedules. It would mean moving from a competitive model to a partnership model (Dwyer et al. 1987; Jae-Nam and Young-Gul 1999; Srinivasan et al. 2011). These partnerships are relationships based on mutual trust, openess, and where the responsibility, authority and decision-making are shared more evenly and there is often an agreement between the parties to share both risks and benefits. (UK Audit Commission 2012; Lambert et al. 1996). In one of a number of reports on the New Zealand red meat sector it was identified that the sector was dominated by commodity supply chains as opposed to differentiated value chains (McLeod et al. 2011). These authors indicated that to address the industry's problems there needed to be greater trust between processors and suppliers and incentives needed to aligned so that one sector did not profit at the expense of the other. There is, currently, little research on what influences farmers to commit to long-term supply chain partnerships. There is significant descriptive research on the characteristics of supply chain partnerships but little explanatory research. This research aims to address this.

New Zealand exports a high proportion of its agri-food products and, despite significant diversification, still relies on a small number of key markets.

China has recently become New Zealand's largest market for dairy products. Over the last 20 years China has moved from being the 31st largest export destination for New Zealand dairy products to the first. This market continues to grow strongly due to rising incomes and urbanisation in China. In contrast, the majority of lamb and venison is exported to the European Union (though China has recently become the largest market for sheep-meat outside of the European Union) (Table 1). Lamb benefits from being counter-seasonal to the European Union domestic supply and 40 per cent is exported by sea freight as chilled cuts. New Zealand has preferential market access for lamb to Europe, with a tariff-free quota of 228,254 tonnes. Venison is supplied into the European Union market primarily in the Northern Hemisphere autumn during the traditional game season, with Germany, the largest single market, taking 40 per cent of total venison exports (Statistics New Zealand 2013). The United States is the main market for New Zealand beef receiving forty per cent of exports with much of it destined for further processing into ground beef.

While dairy production is primarily pasture based there is increasing use of supplementary feeding and irrigation to reduce the impact of climate and to increase production. In contrast, New Zealand meat production is primarily produced on un-irrigated pastures with little use of supplements. This enables low cost, year-round outdoor grazing that produces natural, high quality meat products. It also means that production is highly seasonal with significant variation due to the climate (McLeod et al. 2011; Bensemann et al. 2011). Changes in pasture supply, driven by variations in temperature and rainfall play an important role in supply chain dynamics, affecting price, quality and timing of supply (Bensemann et al. 2011). This is compounded by seasonal and structural overcapacity in the meat processing industry, creating a highly competitive environment for procurement of supply.

Literature review and research framework

The primary objective of strategy is to create a competitive advantage (Barney and Hesterly 2010). Competitive advantage is the ability to produce greater economic value than competing firms (Porter 1985a; Barney and Hesterly 2008, 2010; Lin et al. 1981; Sonka 2003). Porter (1998) identifies three generic strategies firms can use to achieve competitive advantage. The first, a cost leadership strategy, emphasises efficiency and the production of high volumes of standardised products. This provides customers with similar products as competitors but at a lower cost. The second, described as a differentiation strategy, attempts to create products that consumers will pay more for because of attributes they value. The third strategy identifies the breadth of the targeted market segment, where firms attempt to better meet the specific customer needs for a particular market segment. This can involve either a low cost or a differentiated strategy depending of the mature of the market segment.

These generic strategies can also be applied at a supply chain level. Agri-food supply chains have traditionally used a low cost strategy with the provision of large volumes of undifferentiated products and spot market relationships (Sonka 2003). However, many agri-food supply chains are now moving to establish closer relationships with suppliers and customers so they can deliver differentiated products (Hobbs and Young 2001). As consumers demand greater quality and diversity in products and services, buyers need greater commitment from suppliers to ensure a consistent supply of the required quality (Kee-Hung et al. 2005; Fynes et al. 2005).

High levels of commitment mean that suppliers are willing to adapt to meet the required product specifications and committed suppliers will make relationship-specific investments and exert effort to satisfy the buyer (Buxton and Tait 2012). Committed suppliers will allocate the required resources (time, effort and money) to improve their supply chain performance. However, this commitment can also mean suppliers are vulnerable to opportunistic behaviour, especially where they have made relationship-specific investments (Liu 2012). Transaction cost economics identifies the risk of opportunistic behaviour as a determinant of transaction costs. Firms encounter transaction costs as they adopt governance mechanisms to address the risk of opportunistic behaviour (Williamson 1979). Trust is a more effective and lower cost governance mechanism than having formal contracts (Poppo and Zenger 2002; Dyer and Singh 1998; Liu 2012).

This is especially the case when there are complex exchanges requiring co-operation between partners (Poppo and Zenger 2002). Long-term, sustainable partnerships require a high level of collaboration between all parties in the supply chain and are characterised by high levels of trust, commitment, transparency and integrity (Kwon and Suh 2004; Srinivasan et al. 2011). These are also important factors in enabling the efficient and effective flow of information and the allocation of resources in a supply chain (Buxton and Tait 2012). These behaviours are necessary to enable companies to supply differentiated products to customers and achieve a sustainable competitive advantage.

The resource based view (RBV) states that competitive advantage comes from valuable and rare resources, and capabilities. If these are also hard to imitate and not substitutable then they can provide a long-term sustainable competitive advantage (Poppo and Zenger 1997; Barney 1991; Srinivasan et al. 2011). RBV identifies that it is the different resources these firms have that determines the differences in performance between them (Wernerfelt 1984). Examples of the resources are brand names, technical knowledge, skilled human resources, inter-firm relationships, machinery, efficient operating procedures and financial capital. The RBV regards specific assets and, in particular, human assets as being critical to a firm's performance. These provide valuable knowledge and capabilities (Poppo and Zenger 1997). The RBV proposes that companies choose greater integration and more hierarchical governance mechanisms, because with greater investment in specific assets these forms of governance are more efficient (Poppo and Zenger 1997). Originally, the RBV focused only on the resource capabilities located within the individual firm (Barney 1991; Molina and Dyer 1999). However, later developments acknowledged evidence that firms can achieve supply chain productivity gains by making relational investments. Inter-firm relationships enable the combining of resources in unique ways that provide these partnerships with greater competitive advantage. This incorporates the relational exchange perspective into the RBV (Dwyer et al. 1987). This extends the original view of the RBV framework to incorporate intangible resources that exist beyond the boundaries of individual firms (Molina and Dyer 1999).

Firms engage in relationships with other firms to obtain access to complementary resources (Nooteboom et al. 2000). A partner can offer a range of valuable resources, including technical capability, organisational capability, flexibility, reliability, knowledge, innovative capability, network position, international presence and a low risk of discontinuity (Nooteboom 1999). Oliver (1997) suggests that strategic alliances allow firms to obtain assets, competencies or capabilities that cannot be easily purchased in a competitive market for resources. These are, in particular, intangible assets such as specialised technical knowledge, expertise or reputation. Collaboration creates a unique combination of resources that, when combined, have greater value than when on their own. These combinations mean that these resources are more valuable, rare and difficult to imitate (Molina and Dyer 1999). Therefore, long-term supply chain partnerships create a competitive advantage through a number of activities. Partnerships' investment in tangible and intangible relationship-specific assets not only includes things such as specialised machinery, but also includes relational assets such as trust. A significant exchange of knowledge and joint learning can take place that is specific to the relationship. Firms are able to combine scarce resources in complementary ways that enable them to improve quality and efficiency as well as to develop new products and technologies. Through relational governance mechanisms, they are able to lower transaction costs (Molina and Dyer 1999; Dyer and Singh 1998). These create relational rents, which are profits achieved through collaboration that are not able to be produced by each individual firm in isolation.

Social capital theory has become an important perspective within social exchange and social network theory. In incorporating a relational view of social exchange theory, social capital describes the relationship-specific resources that enable the relational rents and is concerned with the nature, structure and resources embedded in a person's network of relationships (Granovetter 1973; Seibert et al. 2001; Burt 1992; Lin et al. 1981). Social capital was initially described by Jacobs (1965), who referred to the networks of community relationships developed over time that provided a basis for trust, co-operation and collective action. Social capital includes the actual and potential resources as a result of relationship networks (Nahapiet and Ghoshal 1998a). Social capital between buyers and suppliers allows them to gain access to, and leverage from, resources residing in their relationships. It reduces the likelihood of conflicts and promotes co-operative behaviour through trust, common goals and social bonds (Villena et al. 2011). Social capital is categorised as cognitive, relational or structural (Nahapiet and Ghoshal 1998b; Villena et al. 2011). Cognitive social capital involves shared visions, goals and culture or, in other words, what you have in common. Structural social capital refers to the overall pattern of connections between actors, in other words, who you have contact with and how you have contact with them (Nahapiet and Ghoshal 1998a). Relational social capital refers to personal relationships of trust, friendship, respect and reciprocity developed through a history of interactions that influences behaviour (Nahapiet and Ghoshal 1998a; Granovetter 1992). Social capital theory is closely aligned with the network view. It assumes that inter-firm relationships are embedded in a network structure (structural social capital), and this affects the behaviour and expectations of firms (Omta et al. 2001). Relational and cognitive social capital describes the characteristics of these network relationships. Many traditional studies of supply chain relationships take a limited linear view and only analyse the dyadic relationships between firms in the supply chain. This approach ignores the complex interdependencies and relationships between firms that exist in a larger supply network (Wilson 2011; Choi and Wu 2009).

This literature review was used to provide a theoretical framework for the research project and shape the interview questions. A resource-based view incorporating social capital theory was the primary lens through which the supplier relationships were viewed. From this it is proposed that suppliers seek to maximise the long-term value of their resources and capabilities. This means they seek to develop and acquire valuable and rare resources and capabilities that are difficult to copy, and this leads to a sustainable competitive advantage. These resources comprise their physical farm resources, which include the soils, topography, climate, location as well as physical structures and buildings. It also refers to their human resources, which include their farm management ability as well as the social capital resources that exist in the relationship with their buyer. Suppliers who are committed to long-term relationships seek to maximise the value of their productive resources by seeking complementary resources in their supply chain partners that can add value to their existing resources as well as create new resources and capabilities. The shared social capital resources are what provide the connections and bonds that facilitate access to these resources.

The main objective of this research is to contribute to the knowledge and understanding of supply chain relationships in the agri-food sector. This will provide a better understanding of how to create long-term committed partnerships between suppliers and buyers in order to meet the higher product specifications and delivery schedules required by international consumers. The research identifies the characteristics of long-term successful supplier/processor/retailer partnerships in New Zealand agri-food supply chains as well as the characteristics of the participants. It identifies how these long-term partnerships create value through co-operation. The research identifies the factors that enable long-term co-operation to occur, as opposed to short-term, opportunistic behaviour and how this co-operative behaviour is maintained.

Methods

The study employed a qualitative case study approach to explore the factors that attract suppliers to be committed to long-term supply relationships in agri-food supply chains. In particular where suppliers are require to consistently deliver to high product specifications. An exploratory case study method was used in order to gain insight into the complex factors that contribute to the formation of long-term supply commitments in agri-food supply chains. Case study research can involve single or multiple cases (Yin 2003). A multiple case study approach was used as this provides advantages in identifying patterns and enables the triangulation of the results.

Semi-structured interviews were undertaken with suppliers from three New Zealand agri-food exporting companies between May 2012 and October 2013. The companies selected all had a focused-differentiation strategy (Porter 1985b) and the products exported included, beef, lamb and venison, and their key markets were in the European Union, North America, Asia and the Middle East. The suppliers were required to meet high product specifications in terms of timing of delivery, food safety, traceability, environmental sustainability, animal welfare and product quality. The suppliers interviewed were located in both the North and South Islands of New Zealand and were from the regions of Canterbury, Otago, Manawatu, Wairarapa, Hawkes Bay and Waikato.

The aim was to understand the characteristics of long term supply chain relationships and the motives of the suppliers who choose to commit to these. The interviewer had a list of questions and topics but an attempted to be led by the supplier in order to ensure the questions didn't limit the scope of the interview and that other important aspects of the supply relationship were not missed. A list of the interview script is provided in Appendix 1. The suppliers were asked what they valued in their relationships with these companies and the benefits they received. The interviews focused on factors such as price and price certainty, relationship quality, benchmarking and information sharing. They were also asked about the costs and risks from supplying these companies.

The producers were asked what they valued in their supply relationships and the benefits they had received. They were also questioned about the disadvantages of supply relationship. The study was exploratory in nature and attempts were made to ensure validity. External validity was achieved through proximity and similarity (in the selection of companies that had similar strategies but different products and markets (Campbell 1986). Internal validity was assured through the number of supplier informants selected within each group while suppliers were selected by the companies involved to provide a broad range of perspectives. They ensured that there were less satisfied suppliers included as well as more contented suppliers.

The case studies were selected to provide perspectives about different companies exporting different products to a range of different markets (Eisenhardt 1989). The criteria for a company's selection was that the company had suppliers who committed to supply on contract with specific product specifications in terms of timing of delivery, food safety, traceability, environmental sustainability, animal welfare and product quality. These suppliers need to keep farm management records and on farm management practices are audited to ensure they meet required animal health and welfare as well as environmental sustainability standards. The suppliers also need to meet specific specifications for things such as the age and weight of the animal and fat cover. The suppliers belonged to "producer groups" where they were had an ongoing supply commitment to the New Zealand exporter. In some cases they produced to requirements of particular retail customer or to specific quality specifications that met the requirements of a number of retail customers. These retail customers often visited the suppliers in New Zealand to communicate with the farmers and understand the farming and production practices in New Zealand.

The companies had to be exporting to high end wholesale or retail customers in the European Union, North America, Asia and the Middle East. The companies were selected to cover beef, lamb and venison export supply chains such that the main New Zealand meat exports were covered. Face-to-face semi-structured on-site interviews were the primary method of data collection. The interviews took between an hour to an hour and a half to complete. A total of 30 suppliers were interviewed from five different producer groups. These were complemented with secondary data such as published company information, supply agreements and newspaper reports. Other secondary data included observations at supplier field days and informal personal communication with suppliers and company personnel. Secondary data provided additional information and validation of the interview data.

Results and discussion

The suppliers from the three companies interviewed had a number of common characteristics that reflected their physical, human and social capital resources and capabilities (Figure 1). The combination of these resources enables these suppliers to develop distinctive competencies. These are unique strengths that enable these suppliers to efficiently deliver reliable supplies of high quality products that meet customer requirements (Hill and Jones 2008). These suppliers choose to commit to these high specification supply chains because it gives them access to complimentary resources, which enables them to maximise the returns on their distinctive competencies. These complimentary resources are the customer relationships, reputation, marketing skills, communication and supplier relationship skills of the companies they supplied.

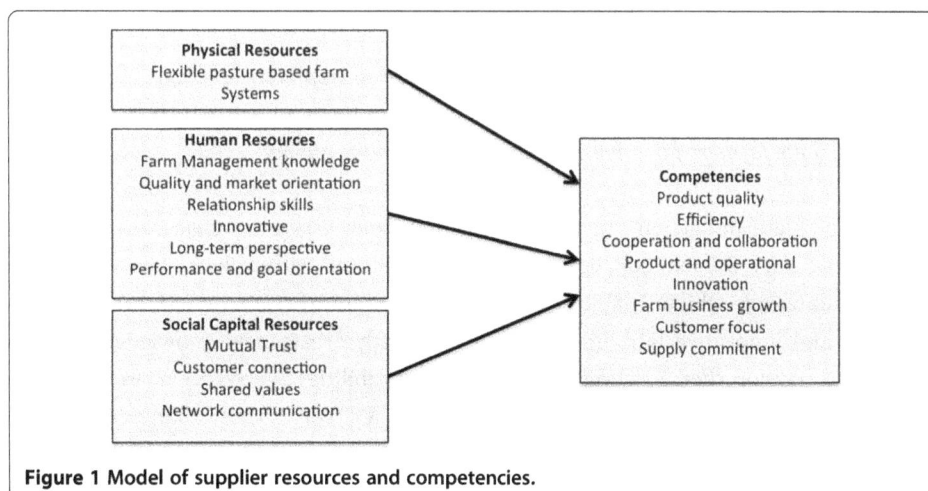

Figure 1 Model of supplier resources and competencies.

The suppliers had farm systems that they could adapt to produce consistently high quality products with more demanding delivery schedules. This involved land and climate resources that enabled a level of production flexibility, or they could achieve this through use of forage crops, irrigation or other stock to balance pasture supply and demand. This was evident when interviewing less committed suppliers as the most common issue they mentioned was the reduced flexibility in delivery timing and quality these supply chains required. This was most significant for suppliers that had farms where summer rainfall was unreliable and soils had little water storage without irrigation.

"We like to be quite flexible and move quite quickly but these things didn't allow us to move as quickly as we would have liked".

"We are a sheep and beef breeding business and our key performance indicator is our ewe production. Trading stock have become a big part of our system so that at any time when its dry, late winter or summer, we can just cut the trading stock".

"Commitment has a cost to it and the reason being that I can't just go and market all my cows as in-calf. Getting involved in this supply chain means we make a commitment that we won't change that policy for the long term and that has a cost. I could sometimes make more money by going to trading".

The human resources and capabilities were a significant factor in the characteristics of these suppliers. They were all capable producers with a high level of farm management ability. Combined with this was a high level of motivation and ability to innovate. They described themselves as progressive farmers, and had a strong desire to develop and grow their farm business. This did not always mean physical expansion but was often about positioning their business to adapt to future changes. As a result they were hungry for information and knowledge that would enable them to improve their farm performance.

"I'd like to see my figures against other suppliers. It's not necessarily to prove I'm the best but just for my satisfaction of seeing where we are and can we improve, and if not, if I'm not up there, then what can I do"

"I think that's what we all need to do. All farmers need to stop being average; It is probably going to be a contradiction of terms. Some farmers farm because that's what they do and some farmers farm because they have to make money".

The desire to create and acquire new knowledge resources was a key characteristic of these suppliers. They valued collaboration and interaction with other like-minded farmers. Collaboration, which enabled the exchange of information and ideas with other capable and innovative farmers, helped them jointly create new knowledge and learning and to develop their existing resources and capabilities. Receiving a premium price for their products was also important. They felt they were "better than average" suppliers and had the capability and resources to produce a high quality product to tighter specifications and, therefore, wanted to receive a reward for this.

"The premium is good but other things are as well. It gives you a pat on the back and know you are doing a good job basically"

"It's important to me to get a premium price and knowing you are doing the right thing to get it"

"The key services for us are providing a good sort of marketing to try and promote high quality beef and sending it to an end market that can pay top dollar for the top product"

"We want to produce a top quality product and a high value with it. We want to know whoever we are moving that product onto is working on the same strategy rather than developing a product and not getting the value out of the market place",

"The premium price is important because we have lots of options here for farming different classes of stock and we pretty much work things back to cents per kg dry matter".

They also had a long-term perspective and wanted to ensure their business was able to adapt to future challenges and changes in the industry. They were goal orientated and motivated by setting both short and long-term goals. The suppliers achieved a great deal of satisfaction from achieving goals and improving performance.

"This year I set a goal at the start of the season and then try to do things as well as I should to achieve that goal. It might seem like a small thing but it's satisfying"

"I definitely take a longer term view maybe the margin should be higher that you're getting but I accept that well that may not be happening now, but it should happen in the future".

"I like the results they give us, the spreadsheet, the benchmarking. I do like that it gives us something to aim for".

The suppliers had strong relationship skills, which enabled them to work cooperatively with other suppliers, and the companies they supplied. They were committed to working co-operatively with other suppliers and other parts of the supply chain. They had learned the benefits of collaboration and working together to create value.

"The thing is you're not competing against anyone; you're not competing on the open market so if you improve your performance it doesn't matter".

"So because you're in this group there's obviously an incentive to actually improve the performance of the whole group".

"There's a strong need for this, a sense of reciprocity, where there's give and take and so I accept that I'm not getting the maximum this year but that's going to pay off in the future. So that's why a key person that's in the group has to be looking long term".

"The whole point of this group is that it's about the good of the group as a whole"

They also had a strong focus on producing high quality products and got a great deal of satisfaction out of this. Many expressed that they were committed to producing high quality products and would do this regardless of the premium received.

"Focusing on quality rather than quantity; if we were focusing on quantity we would be running bulls and trade lambs all over the show. I could make more money by selling all my lambs to sale yards right now rather than having a committed contract, but we don't believe that that is the future of the industry."

They were also customer and market focused. Knowing who the customer was gave them a sense of satisfaction and also gave them assurance that they were adapting their farm system to customer demand; this reduced their perceived risk. Customer connection provided them with personal satisfaction of knowing their efforts to produce a high quality product was appreciated and valued.

"If I left this relationship the customer connection would be one of the main things I would be losing or missing out on."

"The attraction of this supply chain model is you have got a connection with the customer so you can actually see where the money is going and you know the money is all being recycled in the group."

"We like the fact that they are not a normal old beef animal, they are going to a specific market and you are putting the trust in the people who are selling it for you. It is nice to be a little bit more connected to the market of a prime product, which gives us a sense of satisfaction in the quality of what we sell".

"That connection to the customer is really important and that gives you a sense of satisfaction of what you're producing. You know where it's going; it's the whole traceability thing. You know it's going to the top end of the market".

It was clear that these supply chains had significant relationship specific resources. These connections were with the companies they supply, downstream customers as well as other suppliers. These social capital resources enabled suppliers to reduce costs, increase value and reduce risk, which leads to an increase in competitive advantage. A key aspect of the cognitive social capital in the supply chain was shared goals and values. Many suppliers were attracted to these supply chains because they had a common vision with the other supply chain members. This involved producing a high quality product and delivering to customer demands. They wanted to move away from producing commodities and focus on creating more value by meeting customer expectations and being rewarded for doing that.

"I was attracted to this from a marketing point of view; this is the only way we are going to get out of the commodity cycle,"

"We want to produce a top quality product and a high value with it. We want to know that whoever we are moving that product onto is working on the same strategy rather than developing a product and not getting the value out of the market place,"

"Because they're a marketing company that actually aligns with my philosophy over the fact that we should be marketers, not salesmen"

The other members of the supply chain brought complimentary marketing resources that enabled the suppliers to realise a better return from the resources they invested in their farm production. Relational social capital was evident in the strong mutual trust that existed in the supplier–buyer relationships as well as in the horizontal relationships with other suppliers. This was supported by structural social capital with regular interaction and honest communication.

"Totally, totally, I mean, I totally trust all the guys, what they're doing."

Mutual trust and honest communication was also critical as it reduced the risk associated with opportunistic behaviour and enabled them to adapt more quickly to changing market conditions and consumer demands.

"Well one of the things that would damage the relationship would be if they were trying to keep things secret or not telling us. You have to have a fairly good level of trust that they are not hiding any information from you or that they are openly sharing the information that they have."

This social capital was extended through the supply chain to the wholesale and retail customers who often visited suppliers. In some case suppliers had visited the markets and interacted with customers and consumers with in store tastings.

"We were attracted to the scheme because it was not only the price but we knew our meat was going to a specific market – not just disappearing. The Japanese were coming over to and having a look round some farms, which I thought, was good. They took an interest in where the meat came from and made an effort".

Customer connection was important as this provided valuable knowledge exchange and learning. With a greater familiarity about customer needs the suppliers felt they could make strategic investments in their farm production that would create more customer value. The enduring relationships and mutual trust in the supply chain meant that long-term pricing arrangement could be established. The suppliers valued long-term stable prices as this reduced their income volatility. This also enabled improved planning and the ability to invest and focus on maximising production rather than reacting to changing commodity prices. Stable prices gave them better access to financial resources, as banks were more willing to lend if product prices were more predictable.

"You know what the end result is so you can work on margins"

"Having a fixed price is important. You can plan, budget and work towards a good solid outlooks that's consistent. I am not saying they have to have the best price all the time but it's always a big one. As a farmer I can spend the rest of the year planning my crop, changing my rotation to target that"

"What attracted me was the opportunity for a fixed price and focusing on a high value product."

"When I figure out how quickly I can grow them I can go to the bank manager and say that amount of money will come in at that time of year. There is no fluctuation and that for our business going forward is going to be hugely valuable."
"It allows you to focus on improving your farming performance rather than worrying about what the market price is doing".

Relationship quality was important to the suppliers as they sought to establish relationships of mutual trust and reciprocal commitment with their supply chain partners. These aspects of social capital enabled them to mitigate the risk of adapting their production to specific customer requirements and to reduce transaction costs.

"I look after them and they get everything. In return he looks after me and it's a mutual relationship"

Conclusions

The suppliers in the this research confirmed the social capital and resource-based theoretical framework whereby suppliers commit to long-term differentiated supply chains as a strategy to maximise the value of their existing resources and capabilities. They also sought to create opportunities to further develop existing resources through acquiring new resources and capabilities, or to access to complementary assets through their supply chain relationships. This confirmed the resource-based view that firms seek to create a sustainable competitive advantage by controlling valuable and rare resources and capabilities that are difficult to copy. The research also confirmed the social capital perspective as these suppliers saw value in the relational resources that included common goals, mutual trust, communication and social interaction. The suppliers benfited by having long term stable relationships and connection to high value customers. They were able to better customise their production system to meet market demands. This reduced the market risk and also gave them long term stable prices.

The suppliers sought out differentiated supply chains as they identified these as creating greater value for their existing resources. They already have high farm management capabilities as well as quality farm resources so have a greater ability to produce to higher specification products with less flexible delivery requirements. They also have existing social resources in terms of abilities to co-operate and work with others. They have a high level of absorptive capacity, and therefore, can more easily acquire and incorporate new knowledge into their farm systems. The companies they supply provided them with access to complementary resources in the form of access to premium markets where they can achieve greater returns on their resources and capabilities. The social capital resources that existed in the supply chain relationships enabled them to reduce the transaction costs due to their investment in relationship-specific assets.

The case studies showed that it is possible for New Zealand to develop higher value differentiated supply chains with committed long-term relationships. This however, requires a specific set of resources and capabilities that are by definition, rare and difficult to copy. This will only ever be a strategy for a part of the New Zealand agri-food industry. New Zealand needs to develop a diversity of strategies for suppliers and exporters. Individual producers and exporters will choose to supply different types of supply chains within a continuum between spot markets and vertical integration. This will be based on their perception of the way they can maximise the value of their existing resources and capabilities. For example, suppliers with a lower ability to produce or manage consistent quality may maximise their returns by having flexibility in their market arrangements and quality specifications. However, the current industry model is still dominated by commodity supply chains. There is, therefore, a need to specifically support the companies and their suppliers as they were developing these higher value strategies.

The success of differentiated agri-food supply chains requires capable and committed suppliers. This requires significant investment in developing relationships and careful selection of suppliers. Companies developing a differentiated strategy need to identify suppliers who have the ability to produce high quality products and want to be involved in a customer-focused supply chain that provides them with access to premium markets. Companies can build commitment and trust with suppliers through open and transparent communication. They also need to invest in marketing and customer relationships to provide suppliers with access to premium markets so they can be rewarded for the quality of the products they produce.

Although these committed, differentiated supply arrangements will not suit all suppliers, improving the overall resources and capabilities of producers will mean a greater proportion will choose these supply chains as their optimum strategy. This has important implications for policy makers, researchers and for extension services. Private companies, government agencies and industry organisations can support programmes that improve farmer management capability as this will improve the performance of these supply chains as well provide a greater pool of suppliers capable of delivering to these more demanding specifications. New Zealand farm management research has traditionally focused on maximising farm efficiency and reducing costs rather than improving the quality of the product to meet specific customer requirements. More investment needs to be made into research that efficiently adds value rather than on lowering costs.

Farmers need to have both the capability and the motivation to be involved in these supply chains. Many farmers have little awareness of customer demands or opportunities in the market; therefore, promoting knowledge and awareness of market needs and supply chain opportunities is important for providing the understanding and motivation to meet customer needs. Providing resources to improve the physical resources of farms through such things as investment in irrigation systems, improved pasture species and developing enhanced soil quality can improve capability. Providing investment in research and development, and developing farmer knowledge that is specifically targeted at delivering to the specifications of these supply chains, will enable more farmers to have the capability to commit to supplying these customers.

Appendix 1
Interview questions

1. What attracted you to first join the producer group?
2. Why do you think other producers don't join the producer group?

3. What do you think would get more suppliers to join the producer group?

4. What do you value most from being a part of producer group?

5. What do you see as the main benefits of belonging to producer group?

6. How satisfied are you with the performance of your producer group processor/marketer?

7. What do you see as the risks of the being part of the producer group?

8. What do you see as the main costs/disadvantages of belonging to producer group?

9. What do you see as the key services provided by the producer group and how well are these services being delivered today?

10. What would you value that producer group processor/marketer is not currently providing?

Competing interests

The authors declare that they have no competing interests.

References

Audit Commission UK (2012) Working better together. Audit Commission, London

Barney J (1991) Firm resources and sustained competitive advantage. J Manag 17(1):99–120, doi:10.1177/014920639101700108

Barney JB, Hesterly WS (2008) Strategic management and competitive advantage: concepts. Pearson/Prentice Hall, Upper Saddle River, N.J.

Barney JB, Hesterly WS (2010) Strategic management and competitive advantage: concepts and cases. Prentice Hall, Upper Saddle River, N.J.

Bensemann J, Shadbolt N, Conforte D (2011) Farmers' choice of marketing strategies A study of New Zealand lamb producers. Paper presented at the International Food and Agribusiness Conference. IFAMA, Shanghai, China

Burt RS (1992) The social structure of competition. In: Structural Holes. Harvard University Press, Cambridge

Buxton A, Tait J (2012) Measuring fairness in supply chain relationships: Methodology guide. New business models for sustainable trading relationships. International Institute for Environment and Development, Oxfam

Campbell DT (1986) Relabeling internal and external validity for applied social scientists. New Dir Program Eval 1986 (31):67–77

Choi TY, Wu Z (2009) Taking the leap from dyads to triads: Buyer–supplier relationships in supply networks. J Purch Supply Manag 15(4):263–266, doi:10.1016/j.pursup.2009.08.003

Dwyer FR, Schurr PH, Oh S (1987) Developing buyer-seller relationships. J Mark 11:11–27

Dyer JH, Singh H (1998) The relational view: Cooperative strategy and sources of interorganizational competitive advantage. Acad Manag Rev 23(4):660–679

Eisenhardt KM (1989) Building theories from case study research. Acad Manag Rev 14(4):532–550

Fischer C, Hartmann M, Reynolds N, Leat P, Revoredo-Giha C, Henchion M et al (2009) Factors influencing contractual choice and sustainable relationships in European agri-food supply chains. Eur Rev Agric Econ 36(4):541–569, doi:10.1093/erae/jbp041

Fynes B, Voss C, de Búrca S (2005) The impact of supply chain relationship quality on quality performance. Int J Prod Econ 96(3):339–354

Granovetter MS (1973) The strength of weak ties. Am J Sociol 78(6):1360–1380

Granovetter M (1992) Problems of explanation in economic sociology. Netw Organizations: Struct Form Action 25:56

Hill CWL, Jones GR (2008) Strategic Management: An Integrated Approach: An Integrated Approach. Cengage Learning, Boston, MA

Hobbs JE, Young LMC (2001) Vertical linkages in agri-foods supply chains in Canada and the United States. Rev Agric Econ 24(2):428–441

Jacobs J (1965) The death and life of great American cities. Penguin, Harmondsworth

Jae-Nam L, Young-Gul K (1999) Effect of partnership quality on IS outsourcing success: conceptual framework and empirical validation. J Manag Inf Syst 15(4):29–61

Kee-Hung L, Cheng TCE, Yeung ACL (2005) Relationship stability and supplier commitment to quality. Int J Prod Econ 96(3):397–410, doi:10.1016/j.ijpe.2004.07.005

Kwon I, Suh T (2004) Factors affecting the level of trust and commitment in supply chain relationships. J Supply Chain Manag 40(2):4–14, doi:10.1111/j.1745-493X.2004.tb00165.x

Lambert DM, Emmelhainz MA, Gardner JT (1996) Developing and implementing supply chain partnerships. Int J Logist Manag 7(2):1–18

Lin N, Ensel WM, Vaughn JC (1981) Social resources and strength of ties: Structural factors in occupational status attainment. Am Sociol Rev 46(4):393–405

Liu CLE (2012) An investigation of relationship learning in cross-border buyer–supplier relationships: The role of trust. Int Bus Rev 21(3):311–327, doi:10.1016/j.ibusrev.2011.05.005

McLeod A, Mair J, Parker H, Belworthy P (2011) Red meat sector strategy report. Beef & Lamb New Zealand Limited, Auckland

Molina J, Dyer JH (1999) On the relational view: Response to relational view commentary. Acad Manag Rev 24(2):184–186

Nahapiet J, Ghoshal S (1998a) Social capital, intellectual capital, and the organizational advantage. Acad Manag Rev 23(2):242–266, doi:10.5465/amr.1998.533225

Nahapiet J, Ghoshal S (1998b) Social Capital, Intellectual Capital, and the Organizational Advantage. University of Illinois at Urbana-Champaign's Academy for Entrepreneurial Leadership Historical Research Reference in Entrepreneurship

Nelson P (1970) Information and consumer behavior. J Polit Econ 78(2):311–329

Nooteboom B (1999) Inter-firm alliances analysis and design. Routledge, London; New York

Nooteboom B, De Jong G, Vossen RW, Helper S, Sako M (2000) Network interactions and mutual dependence: a test in the car industry. Industry Innov 7(1):117–144, doi:10.1080/713670249

Oliver C (1997) Sustainable competitive advantage: combining institutional and resource-based views. Strateg Manag J 18(9):697–713, doi:10.1002/(sici)1097-0266(199710)18:9<697::aid-smj909>3.0.co;2-c

Omta SWF, Trienekens J, Beers G (2001) Chain and network science: A research framework. J Chain Netw Sci 1(1):

Poppo L, Zenger T (2002) Do formal contracts and relational governance function as substitutes or complements? Strateg Manag J 23(8):707–725

Poppo L, Zenger TR (1997) Testing alternative theories of the firm: Transaction cost, knowledge-based, and measurement explanations for make-or- buy decisions in information services. SSRN eLibrary

Porter ME (1998) Competitive strategy: Techniques for analyzing industries and competitors. Free Press, New York

Porter ME (1985a) Competitive advantage: creating and sustaining superior performance. New York; London: Free Press; Collier Macmillan

Porter ME (1985b) Competitive advantage: Creating and sustaining superior performance. New York: The Free Press: London Collier Macmillan Publishers

Seibert SE, Kraimer ML, Liden RC (2001) A social capital theory of career success. Acad Manag J 1:219–237

Sonka S (2003) Forces driving industrialization of agriculture: implications for the grain industry in the United States. In: Presentation at USDA/ERS Symposium on Product Differentiation and Market Segmentation in Grains and Oilseeds: Implications for an Industry in Transition. pp 27–28

Srinivasan M, Mukherjee D, Gaur AS (2003) Buyer–supplier partnership quality and supply chain performance: Moderating role of risks, and environmental uncertainty. Eur Manag J In Press. Corrected Proof. doi:10.1016/j.emj.2011.02.004

Statistics New Zealand (2013) Global New Zealand – International trade, investment, and travel profile: Year ended December 2013. http://www.stats.govt.nz

Van der Vorst JGAJ (2000) Effective food supply chains; generating, modelling and evaluating supply chain scenarios. PhD. Wageningen University, Wageningen, Netherlands

Villena VH, Revilla E, Choi TY (2011) The dark side of buyer–supplier relationships: A social capital perspective. J Oper Manag 29(6):561–576

Wernerfelt B (1984) A resource-based view of the firm. Strateg Manag J 5(2):171–180, doi:10.1002/smj.4250050207

Williamson O (1979) Transaction-cost economics: The governance of contractual relations. J Law Econ 22(2):233–261

Wilson M (2011) A complex network approach to supply chain network theory. University, Lincoln

Yin RK (2003) Case Study Research Design and Methods. Applied Social Research Methods Series, vol 5, 3rd edn. Sage, Thousand Oaks, California

10

Market impacts of *E. Coli* vaccination in U.S. Feedlot cattle

Glynn T Tonsor[*] and Ted C Schroeder

* Correspondence: gtonsor@ksu.edu
Department of Agriculture
Economics Kansas State University,
342 Waters Hall, Manhattan ITS
66506, USA

Abstract

Immunization through vaccination has been a commercially available pre-harvest intervention to reduce *E. coli* shedding in cattle for about five years. Despite demonstrated substantial improvement in human health that vaccine adoption offers, it has not been widely adopted. This highlights the need for understanding the economic situation underlying limited adoption. Using an equilibrium displacement model, this study identifies the economic impact to U.S. feedlots implementing this vaccination across a series of alternative scenarios. Producers face $1 billion to $1.8 billion in welfare losses over 10 years if they adopt this technology without any associated increases in demand for fed cattle. Retail beef demand increases of 1.7% to 3.0% or export demand increases of 18.1% to 32.6% would each individually make producers economically neutral to adoption. Retail or packer cost decreases of 1.2% to 3.9% would likewise be sufficient to make producers neutral to adoption.

Keywords: Adoption incentive; Beef; Cattle; Cost savings; Demand increases; *E. coli* O157; Economic impacts; Food safety; Vaccination

Shiga toxin producing *E. coli* (STEC O157) is a serious human health hazard in the United States. STEC-related bacteria cause more than 175,000 illnesses (Scallan et al. 2011) with an annual direct economic cost ranging from $489 million (USDA, ERS 2010) to $993 million (Scharff 2010). STEC O157 is naturally occurring in cattle and, through presence in fecal material, threatens food safety if meat contamination occurs during processing.

Because of the human health threat of *E. coli*, considerable beef industry and public health official efforts have targeted pathogen reduction in beef processing plants including development of extensive hazard analysis critical control points (HACCP) and intensive testing of beef for *E. coli* presence (Ferrier and Buzby 2013). Pre-harvest interventions to reduce pathogens in live cattle have arisen as one strategy to lessen chances of post-harvest bacterial contamination of beef. If pathogen presence can be reduced prior to slaughter, the probability of meat contamination during carcass processing will likewise decline (Dodd et al. 2011; Hurd and Malladi 2012).

Vaccines can reduce shedding of *E. coli* in ruminants (Snedeker et al. 2012; Varela et al. 2013; Vogstad et al. 2013). A recently developed commercially available pre-harvest intervention to reduce *E. coli* shedding in cattle is immunization through vaccination (Cull et al. 2012). Despite recognition of the potential reduction in foodborne illness that could result from use of cattle *E. coli* vaccination, adoption is very limited (Callaway et al.

2013; Matthews et al. 2013). Perry et al. (2007) suggest feedlot profits are not directly associated with *E. coli* O157:H7 prevalence as cattle feeding efficiency is not hindered. Furthermore, a well-established market that compensates producers for vaccinating for STEC 0157 has not developed. Thus, an externality exists because feedlots will not implement the socially optimal level of intervention without directly visible economic incentives. Doing so adds costs without directly visible offsetting increases in revenue. The fact that producers do not have direct incentives to employ *E. coli* vaccination, even though doing so would increase beef food safety, led a recent USA Today article to claim "the economics are backwards" (Weise 2011).

Despite the obvious importance to food safety and human foodborne illness, the economic feasibility and impacts of producer adoption of cattle immunization against *E. coli* have not been determined. The purpose of this study is to evaluate the economic impacts of incorporating animal vaccination into *E. coli* pre-harvest control practices. Specifically this study estimates direct producer costs associated with use of a vaccine in cattle feeding, referred to here as an *E. coli* vaccine. Direct costs include vaccine cost, costs of administering, and potential animal feeding performance impacts associated with the vaccination. Potential benefits include reduced packer or retailer costs associated with lower risk of pathogens, reduced food safety concerns, and potentially increased domestic consumer or export demand associated with safer beef. To estimate market level impacts of the vaccination, we use an equilibrium displacement model (EDM) that incorporates supply and demand shifts associated with the cattle immunization to determine economic impacts of the food safety technology across a series of alternative scenarios.

Estimating economic impacts of *E. coli* vaccines being adopted by feedlot operators is important for several reasons. First, feedlot operators need additional information to make sound adoption decisions. Secondly, understanding broader market impacts of possible adoption highlights how net benefits are distributed throughout the industry. Third, society's ongoing interest in food safety and associated desire for regular improvements in risk mitigation protocols further motivates interest from those outside the beef production chain. Given the apparent market failure of *E. coli* vaccination adoption, an assessment of economic impacts and sensitivity to various market reactions has important policy implications.

Background

Vaccination against *E. coli* O157:H7 and fecal shedding has been available in the United States for over five years with the first licensed vaccine approved in February 2009. The vaccine is a siderophore receptor and porin (SRP[*]) protein exclusively marketed by Zoetis[a]. The vaccine is administered to feedlot cattle during the feeding phase of production with two or three doses[b].

A couple of particularly noteworthy studies have examined the effectiveness of the vaccine in reducing fed cattle fecal concentrations of *E. coli* O157:H7 and in its impact on cattle feeding performance. Thomson et al. (2009) found use of the SRP[*] vaccine reduced fecal shedding concentrations in fed cattle by up to 98% and cattle feeding performance was unaffected. Cull et al. (2012) found that two doses of SRP[*] reduced shedding by more than 50% and reduced high shedders by more than 75%. However, Cull et al. (2012) identified a small, but statistically and economically important, reduction in cattle feeding performance associated with vaccinating. In particular, average daily gain declined by 2.7% and feed conversion increased by 2.1% for vaccinated relative to unvaccinated cattle. The reduction in animal performance was hypothesized to be associated with the second vaccination where the vaccinated animals were processed in a chute an additional time relative to the control non-vaccinated cattle. In Thomson et al. (2009) the control cattle were vaccinated with a placebo each time, so the numbers of chute processes were the same for the control and *E. coli* vaccinated cattle. Running cattle through a chute can result in cattle shrink that may not be fully recaptured when placed back on feed and can temporarily disrupt cattle feed intake (Blasi et al. 2009).

Given the differences in experimental results relative to cattle feeding performance impacts associated with administration of an *E. coli* vaccine, we allow for alternative assumptions in our economic analyses. In particular, we consider two alternatives, one assuming no reduction in animal performance and a second assuming a 2.7% reduction in average daily gain and a 2.1% increase in feed conversion (pounds of feed per pound of gain) as a result of the vaccination.

Important to also consider are possible demand improvements or cost savings that could potentially follow implementation of *E. coli* vaccine programs at the feedlot level. Moghadam et al. (2013) estimate that *E. coli* O157:H7 recalls by the USDA Food Safety Inspection Service (FSIS) have a rapid and important economic impact on live cattle futures markets amounting to approximately $6/head for all cattle slaughtered in the United States. Domestic beef demand has been harmed by past FSIS recall events (Marsh et al. 2004; Piggott and Marsh 2004; Tonsor et al. 2010). Furthermore, export market demand for U.S. beef is highly sensitive to food safety (Bailey 2007). As such, domestic retail beef demand and wholesale export beef demand could improve with *E. coli* vaccine programs that reduce *E. coli* prevalence.

Extensive research has examined the cost impacts of additional food safety programs and protocols being introduced into the U.S. beef industry (Antle 1999; 2000). Costs include direct production costs such as additional labor requirements, slowing down processing line speed, investing in food safety technologies, modifying processing procedures or facilities, and expenses of more intensive product sampling and food safety testing (Ferrier and Buzby 2013). In face of a recall, costs of plant down-time, clean up, physical product losses, costs of completing a food recall, and loss of firm customers and reputation can collectively be substantial. Important expenses also include possible litigation costs associated with foodborne human illnesses that have proved expensive in cases where meat food safety breaches have occurred (e.g., Gabbett 2010; Scott 2012).

Given the sizeable costs involved to downstream firms in light of a food safety event, substantial incentives are present to reduce the probabilities of such events including the possibility of feedlots implementing *E. coli* vaccine programs. In essence, use of an *E. coli* vaccine by cattle producers could result in notable benefits to downstream firms but direct benefits to producers that incur the costs of adoption are currently elusive, limiting adoption.

Given the economic importance of identifying the economic impact *E. coli* vaccine program introduction could have on stakeholders throughout the meat and livestock chain we first directly estimate market impacts in the absence of incentivizing demand improvements or possible downstream cost savings. Given the unknown nature of possible demand or cost improvements we then proceed to estimate market impacts under alternative market scenarios to identify the specific beef demand improvements or cost savings that would be needed to make feedlot operators in aggregate indifferent to adoption.

Estimating and discussing these demand or cost improvements that may lead to feedlot adoption is important for several reasons. First, as the levels of demand or cost adjustments that could be experienced elsewhere in the supply chain are realized, having critical thresholds identified is valuable. Second, cattle producers would adopt vaccine programs if they received offsetting benefits which reinforces the value in identifying necessary demand or cost improvements further upstream to encourage adoption (Smith et al. 2013). Finally, from a policy perspective, any cost-benefit analysis of alternative beef food safety enhancing interventions requires the information provided in this study regarding market impacts of these interventions.

Methods

The methodological approach can succinctly be described as using estimates of costs increases incurred by feedlots implementing *E. coli* vaccination programs and estimating changes in prices and quantities at market levels spanning the vertically linked beef industry as well as connected pork and poultry markets. The initial exogenous market shock is feedlot level production costs increasing leading to an inward shift in fed cattle supply. To estimate the market level impact of *E. coli* vaccination on prices and quantities throughout the livestock and meat industry we employ an equilibrium displacement model (EDM). The EDM utilized here is similar to that used by Schroeder and Tonsor (2011) and is documented in the appendix.

The EDM is composed of four sectors in the beef industry: 1) retail (consumer), 2) wholesale (processor/packer), 3) fed cattle (cattle feeding in feedlots), and 4) farm (feeder cattle from cow-calf producers). To capture interactions between retail meat substitutes for beef we also include the pork and poultry markets. Reflecting the higher degree of integration relative to the beef industry, the economic model includes three pork marketing chain sectors (retail, wholesale, and fed cattle) and the poultry marketing chain is composed of two sectors (retail and wholesale). International trade is explicitly incorporated in the model at the wholesale level for all three species. The resulting framework is consistent with existing research and follows the recent work of Brester et al. (2004) and Pendell et al. (2010).

Given estimated changes in prices and quantities, producer surplus at the feedlot level where the adoption decision occurs, as well as other levels in the vertically linked supply chain, is calculated as a widely accepted measure of economic welfare impact. As in most EDM applications, direct estimation of elasticities is prohibitive because of the large number of equations and identification problems in jointly estimating supply and demand relationships (Brester et al. 2004). However, given the *E. coli* vaccination results in relatively small aggregate market shifts (in proportional terms), we follow standard EDM procedures and utilize elasticity estimates reported in the published literature to parameterize the model.

We simulate our model annually for ten consecutive years to trace a hypothetical adoption path over time by producers of the *E. coli* vaccination technology. Consistent with historical beef cattle cycles, we assume it takes the marketplace ten years to fully adjust from short-run to long-run relationships. Ten years of market effects were simulated by linearly adjusting all elasticities between short-run (year 1) and long-run (year 10) using elasticity estimates employed by Pendell et al. (2010)[c]. Supply, demand, and quantity transmission elasticities used are equivalent to those used by Schroeder and Tonsor (2011). Similarly, base price and quantity values are necessary to estimate surplus calculations. The market price and quantity values are annual averages for calendar year 2012 as reported by the Livestock Marketing Information Center (LMIC).

Our analysis assumes 10% of fed cattle are vaccinated in year 1, 25% in year 2, 50% in year 3, and 90% in years 4–10. This reflects a typical "S-curve" adoption pattern where adoption increases rapidly upon introduction of the technology with a plateau corresponding to the fact that few technologies are ever completely adopted by all parties in a heterogeneous industry. The employed adoption rate would of course only occur if private market incentives to adopt were widely present and accessible to producers, which currently they are not. As such, the adoption rate is used here for exemplification and estimation of cost impacts and is not a forecast of a probable adoption path given current market conditions.

Results

When feedlot operators implement *E. coli* vaccination protocols one main direct impact serves as the initial shock in our EDM. Specifically, production costs increase leading to an inward shift in fed cattle supply. This initial exogenous shock initiates a ripple-effect through the industry as reflected in multiple endogenous shifts outlined within the EDM.

The change in net returns of finishing cattle for those implementing *E. coli* vaccinations were calculated under alternative assumptions following Lueger et al. (2012). When assuming no animal performance impact, the direct costs are estimated at $6.47/head with the vaccination being administered twice to each animal. The first vaccination occurs upon arrival at the feedlot where cattle are all processed through a chute anyway (so no additional chute charge or animal processing is associated uniquely to the *E. coli* vaccination). For the second vaccination, additional chute and labor charges occur since cattle would not generally be processed through the chute again unless being vaccinated for *E. coli*. One potentially important addition to the direct costs of vaccinating is whether an adverse animal performance outcome occurs from the vaccination. Research includes no impact (Thomson et al. 2009) to an observed animal performance decline (Cull et al. 2012). As such, a second cost scenario is assumed where the direct costs that include vaccination and animal performance losses are estimated at $13.11/head (Lueger et al. 2012). In addition to these base direct costs, adopting feedyards could incur costs associated with third-party verification that *E. coli* vaccinations were indeed implemented if packers were going to pay them for such a verified production protocol. We assume *E. coli* vaccination verification costs of $1.88/head which are based on costs for age and source verification identified by Pendell et al. (2013).

Given the magnitude of the direct costs for feedlots to vaccinate, they generally will not without a clear direct economic incentive. As such, determining the downstream benefits that would need to occur to encourage adoption is essential to understand if adoption is desirable. The $/head implementation costs are presented in Table 1 along with the exogenous supply shifts these cost increases represent in each year within the EDM given an average fed cattle value in 2012 of $1,604/head.

Table 1 Exogenous fed cattle production cost increase of vaccination and verification program adoption

Scenario:	Direct Cost	Percentage inward supply shift			
		Year 1	Year 2	Year 3	Years 4-10
		10%	25%	50%	90%
	$/hd[a]	Adoption	Adoption	Adoption	Adoption
No animal performance impact	$8.35	0.052%	0.130%	0.260%	0.468%
Animal performance impact	$14.99	0.093%	0.234%	0.467%	0.841%

[a]Includes $6.47 per head cost of vaccinating plus $1.88 per head verification cost under no animal performance impact and $13.11 per head cost under performance impact plus $1.88 per head verification cost.

The EDM was first applied to identify economic impacts in the case of *E. coli* vaccination program implementation without additional demand or cost benefits occurring elsewhere in the supply chain. This scenario of course would not happen because feedlots will not adopt without clear direct benefits. But having this scenario is necessary to determine the subsequent magnitudes of market events that would need to occur to entice adoption.

Table 2 summarizes the short- (1-year), intermediate- (5-year), and long-run (10-year) changes in prices and quantities estimated by the EDM where animal performance impacts are omitted and considered. Retail and wholesale beef as well as fed cattle, and feeder cattle quantities all decline in each of the 10 years considered. The reduced production volumes reflect the inward shift in fed cattle supply, derived inward shifts in wholesale and retail beef supplies, and derived demand reductions experienced at the feeder cattle level following the increased production costs for feedlots. Retail and wholesale beef prices increase in all 10 years as the increased feedlot costs pass vertically towards consumers. Feeder cattle prices decline in all 10 years as feedlot vaccination costs reduce derived demand for feeder cattle. The fed cattle price path switches signs over the 10 years examined. Specifically, fed cattle prices decline in years 1 and 2 and increase over years 3–10 reflecting long-run supply being more elastic than short-run supply and a multitude of derived demand and supply feedbacks captured by the model. The quantities of wholesale beef exported and imported, as well as the price of imported wholesale beef all decline over the 10 years evaluated. This primarily follows an overall reduction in wholesale beef supplies and increased wholesale beef prices. More broadly, the long-run impacts are smaller as the entire supply chain adjusts to *E. coli* vaccination program implementation over time.

In the base situation of no additional demand enhancement or cost savings, the cumulative net present value producer surplus losses over ten years at the feedlot level are $1.00 billion if no animal performance reduction occurred by vaccinating and $1.79 billion if reduced animal performance is considered (Table 3). This substantial difference in welfare, despite small potential impacts on animal performance, clearly illustrates how the economic value of interventions changes if animal productivity is affected.

These substantial losses reflect changes in prices and quantities summarized in Table 2 and occur if the adoption rate we assume and no offsetting benefits materialize. This illustrates why, consistent with Matthews et al. (2013) and Callaway et al. (2013), limited voluntary adoption of *E. coli* vaccination will occur unless recognized direct incentives for implementation arise. Such incentives could occur in the form of derived demand increasing for fed cattle from feedlots with *E. coli* vaccination programs if either domestic retail or wholesale export beef demand increased following program

Table 2 Percentage change in endogenous variables of the equilibrium displacement models with adoption costs but no benefits scenario

Endogenous variables	No animal performance impact			With animal performance impact		
	Short run	Intermediate	Long run	Short run	Intermediate	Long run
Retail beef quantity	−0.27%	−0.04%	−0.01%'	−0.49%	−0.06%	−0.01%
Retail beef price	0.32%	0.04%	0.01%	0.58%	0.06%	0.01%
Wholesale beef quantity	−0.51%	−0.15%	−0.04%	−0.91%	−0.27%	−0.07%
Wholesale beef price	0.40%	0.16%	0.04%	0.72%	0.28%	0.07%
Slaughter cattle quantity	−0.40%	−0.28%	−0.12%	−0.72%	−0.51%	−0.21%
Slaughter cattle price	−0.38%	0.26%	0.14%	−0.68%	0.46%	0.25%
Feeder cattle quantity	−0.23%	−0.21%	−0.09%	−0.42%	−0.37%	−0.16%
Feeder cattle price	−1.06%	−0.15%	−0.03%	−1.91%	−0.27%	−0.06%
Imported wholesale beef quantity	−0.38%	−0.13%	−0.04%	−0.69%	−0.24%	−0.07%
Exported wholesale beef quantity	−0.17%	−0.24%	−0.11%	−0.30%	−0.44%	−0.20%
Imported wholesale beef price	−0.21%	−0.02%	0.00%	−0.37%	−0.04%	−0.01%
Retail pork quantity	0.04%	0.01%	0.00%	0.08%	0.01%	0.00%
Retail pork price	0.02%	0.00%	0.00%	0.04%	0.00%	0.00%
Wholesale pork quantity	0.03%	0.01%	0.00%	0.05%	0.01%	0.00%
Wholesale pork price	0.02%	0.00%	0.00%	0.03%	0.00%	0.00%
Slaughter hogs quantity	0.01%	0.00%	0.00%	0.02%	0.01%	0.00%
Slaughter hogs price	0.03%	0.00%	0.00%	0.06%	0.01%	0.00%
Imported wholesale pork quantity	0.02%	0.00%	0.00%	0.03%	0.01%	0.00%
Exported wholesale pork quantity	−0.02%	0.00%	0.00%	−0.03%	0.00%	0.00%
Imported wholesale pork price	0.01%	0.00%	0.00%	0.02%	0.00%	0.00%
Retail poultry quantity	0.05%	0.01%	0.00%	0.08%	0.01%	0.00%
Retail poultry price	0.04%	0.00%	0.00%	0.08%	0.00%	0.00%
Wholesale poultry quantity	0.05%	0.01%	0.00%	0.09%	0.01%	0.00%
Wholesale poultry price	0.00%	0.00%	0.00%	0.00%	0.00%	0.00%
Exported wholesale poultry quantity	−0.32%	−0.04%	−0.01%	−0.57%	−0.08%	−0.01%

Note: These percentage changes are relative to 0% vaccination and verification adoption. Short-, intermediate-, and long-run corresponds to years 1, 5, and 10, respectively from the EDM simulated over 10 consecutive years. Percentage changes for each individual year are available upon request.

implementation. Similarly, a derived demand benefit could materialize if production costs at either the retail or wholesale level declined following *E. coli* vaccination program implementation. Given these unknown but important plausible alternatives, we extended our analysis and utilized the EDM to identify demand benefits or cost savings needed to make the feedlot level indifferent to adoption.

Table 4 presents the estimated retail demand increase, wholesale export demand increase, retail cost savings, and wholesale costs savings that result in no changes in producer surplus at the fed cattle (feedlot) level[d]. Note, the estimates in Table 4 are for independent downstream shocks to demand or costs needed to make feedlot producers economically indifferent to adoption. Possible combinations of demand and cost impacts would be smaller than individual shocks necessary to make feedlot producers indifferent. The minimum changes that may lead to feedlot adoption are lower in each case where implementing an *E. coli* vaccination program does not impact animal performance.

Table 3 Producer surplus changes from vaccination and verification program adoption ($ millions), no benefits scenario

Beef producer surplus	No animal performance impact				With animal performance impact			
	Short run	Intermediate run	Long run	Cumulative present value	Short run	Intermediate run	Long run	Cumulative present value
Retail level	294.05	32.58	5.07	654.66	527.41	58.49	9.10	1,174.59
Wholesale level	197.81	76.67	18.49	804.03	354.46	137.58	33.20	1,441.69
Fed cattle level	−224.72	−110.56	−173.92	−1,000.90	−402.86	−198.29	−312.13	−1,795.23
Feeder cattle level	−434.83	−61.77	−12.98	−1,018.27	−780.04	−110.82	−23.30	−1,826.42
Total Beef Industry Producer Surplus	−167.69	−63.08	−163.33	−560.48	−301.02	−113.04	−293.13	−1,005.37

Note: Producer surplus is calculated relative to 2012 quantities and prices for livestock and meat. Cumulative net present value was calculated using a 5% discount rate.

Table 4 Independent changes needed for no change in feedlot sector producer surplus

	No animal performance impact	With animal performance impact
Domestic retail beef demand increase	1.7%	3.0%
Wholesale export beef demand increase	18.1%	32.6%
Retail beef (retailer) cost decrease	2.2%	3.9%
Wholesale beef (packer) cost decrease	1.2%	2.2%

Note: Values are the exogenous responses (demand increases or cost savings) resulting in cumulative net present value of producer surplus not changing at the fed cattle level (feedlots). Demand increases on all beef production were considered while cost savings evaluated corresponded only to the portion of product retailers or packers would receive from *E. coli* vaccination programs reflecting the 10-year adoption path assumed.

When animal performance impacts are considered, either a 3.0% increase in domestic retail beef demand or a 32.6% increase in wholesale export beef demand would provide the derived demand benefits to make feedlot operators indifferent to implementation. The 3.0% domestic retail demand increase is within the range of experienced annual demand shifts in the U.S. (AgManager 2013). However, the finding of Zingg and Siegrist (2012) that a minority of consumers may be willing to consume meat from vaccinated animals casts some doubt on the extent of a positive, aggregate retail demand response. In 2012 Japan and South Korea combined accounted for 30.8% of total U.S. beef exports (LMIC 2013). Accordingly, context on the 32.6% wholesale export increase can readily be made noting how maintaining access to Japan and South Korea by avoiding food-safety related market closures could offset feedlot level vaccination program costs. The identification of thresholds for export demand increases being approximately 10 times those of domestic demand increases reflects the fact approximately 90% of beef produced in the U.S. is consumed domestically. While domestic and export demand responses are unknown, recognizing the demand response thresholds is important for broader industry-wide deliberations and sets the stage for additional future research.

Table 4 also indicates that 3.9% (2.2%) cost savings for retailers or 2.2% (1.2%) cost savings for packers results in no net economic welfare changes for the feedlot segment if animal productivity is (not) reduced through a vaccination protocol. As with the threshold demand values, the precise level of cost savings at retail or wholesale levels are unknown as adjustments that operations may make in *E. coli* mitigation efforts have yet to be directly studied. Hurd and Malladi (2012) concluded that the number of ground beef-related *E. coli* human illnesses in the United States could be reduced by 58% from about 20,000 illnesses to around 8,400 per year under an 80% effective and fully adopted feedlot steer and heifer vaccination program. Smith et al. (2013) found combinations of interventions applied pre-harvest and throughout processing resulted in larger relative *E. coli* risk reductions. If a single and relatively simple intervention such as a vaccination program would have this dramatic of impact on foodborne illnesses, downstream cost savings would certainly be realized. Estimating potential downstream cost savings is an area ripe for future research.

Discussion and Conclusions

This study expands knowledge of economic implications of implementing *E. coli* vaccination programs, highlights key areas of where additional research would be valuable, and provides information that could improve societal response to efficiently mitigating food safety risks. Vaccinations by feedlots could potentially reduce *E. coli* related human foodborne illnesses from ground-beef by 58% (Hurd and Malladi 2012). However, cattle producers will not adopt *E. coli* vaccination programs without offsetting direct benefits because doing so is costly and would reduce their economic welfare by $1 billion to $1.8 billion. Currently, direct benefits of *E. coli* vaccine adoption by cattle producers are elusive as a well-established market premium does not exist despite the vaccine having been commercially available for over five years now.

What might it take for *E. coli* vaccination programs to be successfully implemented? We illustrate the threshold magnitudes of downstream demand improvements or cost savings that are needed to provide producers economic incentives to adopt *E. coli* vaccination programs. Domestic consumer demand increases of 2-3%, export wholesale market increases of 18-33%, retailer cost reductions of 2-4%, or processor cost reductions of 1-2% would each individually be sufficient to make producer adoption welfare neutral.

This study also highlights the need to further examine if human health benefits from implementing *E. coli* vaccination programs are significant enough to consider additional policy adjustments that encourage adoption. This issue is beyond the economics of adoption focused scope of this paper but certainly is an area of importance for future research. Similarly, a valuable area for future research is to consider the demand increase and cost reduction thresholds identified in this study and determine whether and how *E. coli* vaccination adoption incentives might occur and translate incentives back to cattle producers.

Endnotes

[a]See: https://online.zoetis.com/US/EN/Solutions/Pages/SRPEcoli/index.aspx

[b]Experiments with both two doses and three doses have been conducted (Cull et al. 2012; Thomson et al. 2009) with three doses showing a trend in efficacy.

[c]Available at: http://ajae.oxfordjournals.org/content/suppl/2010/04/29/aaq037.DC1/aaq037supp.pdf

[d]Demand increases were modelled to impact 100% of production over the 10 year period while cost increases were modelled to impact only the portion of production that aligns with feedlot vaccination adoption. That is, demand increases reflect an assumption of product being undifferentiated downstream to buyers while cost increases reflect an assumption of downstream purchases only experiencing cost savings when product is verified to be sourced from adopting feedyards.

[e]This model follows Pendell et al. (2010) and similar studies in assuming international trade can be succinctly captured by including exchange of meat products while not explicitly incorporating live animal trade.

[f]Available at: http://ajae.oxfordjournals.org/content/suppl/2010/04/29/aaq037.DC1/aaq037supp.pdf

APPENDIX - Details of Applied Equilibrium Displacement Model

To estimate the market level impact of *E. coli* vaccination we employ an equilibrium displacement model (EDM). The EDM utilized here is similar to that used by Schroeder and Tonsor (2011). The EDM is composed of four sectors in the beef industry: 1) retail (consumer), 2) wholesale (processor/packer), 3) fed cattle (cattle feeding in feedlots), and 4) farm (feeder cattle from cow-calf producers). To capture interactions between retail meat substitutes for beef we also include the pork and poultry markets. Reflecting the higher degree of integration relative to the beef industry, the economic model includes three pork marketing chain sectors (retail, wholesale, and fed cattle) and the poultry marketing chain is composed of two sectors (retail and wholesale). International trade is explicitly incorporated in the model at the wholesale level for all three species. The resulting framework is consistent with existing research and most closely follows the recent work of Brester et al. (2004) and Pendell et al. (2010). The structural model (omitting error terms for convenience) is given by the following series of general demand and supply equations of this multi-species model. Superscripts r, w, s, and f denote the retail, wholesale, fed cattle, and farm market levels, respectively; subscripts B, K, and Y denote beef, pork, and poultry, respectively; P is price; Q is quantity; and Z and W denote demand and supply shifters, respectively. Consistent with existing international trade, the model captures imports (subscript i) and exports (subscript e) of beef, pork, and poultry[e]. Equations (1) - (25) omit superscripts for demand and supply as market clearing conditions are imposed, requiring demand and supply to equal.

Beef marketing chain

(1) Retail beef primary demand: $Q_B^r = f_1\left(P_B^r, P_K^r, P_Y^r, Z_B^r\right)$,

(2) Retail beef derived supply: $Q_B^r = f_2\left(P_B^r, Q_B^w, W_B^r\right)$,

(3) Wholesale beef derived demand: $Q_B^w = f_3\left(P_B^w, Q_B^r, Z_B^w\right)$,

(4) Wholesale beef derived supply: $Q_B^w = f_4\left(P_B^w, Q_B^s, Q_{Bi}^w, Q_{Be}^w, W_B^w\right)$,

(5) Imported wholesale beef derived demand: $Q_{Bi}^w = f_5\left(P_{Bi}^w, Q_B^w, Z_{Bi}^w\right)$,

(6) Imported wholesale beef derived supply: $Q_{Bi}^w = f_6\left(P_{Bi}^w, W_{Bi}^w\right)$,

(7) Exported wholesale beef derived demand: $Q_{Be}^w = f_7\left(P_B^w, Z_{Be}^w\right)$,

(8) Fed cattle derived demand: $Q_B^s = f_8\left(P_B^s, Q_B^w, Z_B^s\right)$,

(9) Fed cattle derived supply: $Q_B^s = f_9\left(P_B^s, Q_B^f, W_B^s\right)$,

(10) Farm (feeder cattle) derived demand: $Q_B^f = f_{10}\left(P_B^f, Q_B^s, Z_B^f\right)$,

(11) Farm (feeder cattle) primary supply: $Q_B^f = f_{11}\left(P_B^f, W_B^f\right)$,

Pork marketing chain

(12) Retail pork primary demand: $Q_K^r = f_{12}\left(P_B^r, P_K^r, P_Y^r, Z_K^r\right)$,

(13) Retail pork derived supply: $Q_K^r = f_{13}\left(P_K^r, Q_K^w, W_K^r\right)$,

(14) Wholesale pork derived demand: $Q_K^w = f_{14}\left(P_K^w, Q_K^r, Z_K^w\right)$,

(15) Wholesale pork derived supply: $Q_K^w = f_{15}\left(P_K^w, Q_K^s, Q_{Ki}^w, Q_{Ke}^w, W_B^w\right)$,

(16) Imported wholesale pork derived demand: $Q_{Ki}^w = f_{16}\left(P_{Ki}^w, Q_K^w, Z_{Ki}^w\right)$,

(17) Imported wholesale pork derived supply: $Q_{Ki}^w = f_{17}\left(P_{Ki}^w, W_{Ki}^w\right)$,

(18) Exported wholesale pork derived demand: $Q_{Ke}^w = f_{18}\left(P_K^w, Z_{Ke}^w\right)$,

(19) Market hog derived demand: $Q_K^s = f_{19}\left(P_K^s, Q_K^w, Z_K^s\right)$,

(20) Market hog primary supply: $Q_K^s = f_{20}\left(P_K^s, W_K^s\right)$,

Poultry marketing chain

(21) Retail poultry primary demand: $Q_Y^r = f_{21}\left(P_B^r, P_K^r, P_Y^r, Z_Y^r\right)$,

(22) Retail poultry derived supply: $Q_Y^r = f_{22}\left(P_Y^r, Q_Y^w, Q_{Ye}^r, W_Y^r\right)$,

(23) Wholesale poultry derived demand: $Q_Y^w = f_{23}\left(P_Y^w, Q_Y^r, Z_Y^w\right)$,

(24) Wholesale poultry primary supply: $Q_Y^w = f_{24}(P_Y^w, W_Y^w)$,

(25) Exported wholesale poultry derived demand: $Q_{Ye}^w = f_{25}(P_Y^w, Z_{Ye}^w)$.

Consistent with Wohlgenant (1993), we incorporate variable input proportions by allowing production quantities to vary across the market levels in the marketing chain. Totally differentiating equations (1) - (25), including variable input proportions, and placing all the endogenous variables on the left-hand side of each equation and isolating exogenous effects to the right-hand side of each equation results in the following EDM. E represents a relative change operator (i.e., $EQ = d \ln Q = dQ/Q$); η_a^m is the own-price elasticity of meat/species a demand at market level m; η_{ab}^m is the cross-price elasticity of demand for meat a with respect to retail prices of meat b; ε_a^m is the own-price elasticity of meat/species a supply at market level m; τ^{lm} is the percentage change in quantity demanded at market level m given a 1% change in quantity demanded at market level l; γ^{lm} is the percentage change in quantity supplied at market level m given a 1% change in quantity supplied at market level l. In this specification, market levels are linked by downstream quantity variables among the demand equations and upstream quantity variables among the supply equations (Wohlgenant 1993).

Beef marketing chain

(1″) Retail beef primary demand: $EQ_B^r - \eta_B^r EP_B^r - \eta_{BK}^r EP_K^r - \eta_{BY}^r EP_Y^r = EZ_B^r$,

(2″) Retail beef derived supply: $EQ_B^r - \varepsilon_B^r EP_B^r - \gamma_B^{wr} EQ_B^w = EW_B^r$,

(3″) Wholesale beef derived demand: $EQ_B^w - \eta_B^w EP_B^w - \tau_B^{rw} EQ_B^r = EZ_B^w$,

(4″) Wholesale beef derived supply:

$$EQ_B^w - \varepsilon_B^w EP_B^w - \gamma_B^{sw}(Q_B^s/Q_B^w)EQ_B^s - (Q_{Bi}^w/Q_B^w)EQ_{Bi}^w + (Q_{Be}^w/Q_B^w)EQ_{Be}^w = EW_B^w,$$

(5″) Imported wholesale beef derived demand: $EQ_{Bi}^w - \eta_{Bi}^w EP_{Bi}^w - \tau_B^{rw} EQ_B^w = (Q_{Bi}^w/Q_B^w)EZ_{Be}^w + EZ_{Bi}^w$,

(6″) Imported wholesale beef derived supply: $EQ_{Bi}^w - \varepsilon_{Bi}^w EP_{Bi}^w = EW_{Bi}^w$,

(7″) Exported wholesale beef derived demand: $EQ_{Be}^w - \eta_{Be}^w EP_B^w = EZ_{Be}^w$,

(8″) Fed cattle derived demand: $EQ_B^s - \eta_B^s EP_B^s - \tau_B^{ws} EQ_B^w = (Q_{Be}^w/Q_B^w)EZ_{Be}^w + EZ_B^s$,

(9″) Fed cattle derived supply: $EQ_B^s - \varepsilon_B^s EP_B^s - \gamma_B^{fs} EQ_B^f = EW_B^s$,

(10″) Farm (feeder cattle) derived demand: $EQ_B^f - \eta_B^f EP_B^f - \tau_B^{sf} EQ_B^s = EZ_B^f$,

(11″) Farm (feeder cattle) primary supply: $EQ_B^f - \varepsilon_B^f EP_B^f = EW_B^f$,

Pork marketing chain

(12″) Retail pork primary demand: $EQ_K^r - \eta_{KB}^r EP_B^r - \eta_K^r EP_K^r - \eta_{KY}^r EP_Y^r = EZ_K^r$,

(13″) Retail pork derived supply: $EQ_K^r - \varepsilon_K^r EP_K^r - \gamma_K^{wr} EQ_K^w = EW_K^r$,

(14″) Wholesale pork derived demand: $EQ_K^w - \eta_K^w EP_K^w - \tau_K^{rw} EQ_K^r = EZ_K^w$,

(15″) Wholesale pork derived supply: $EQ_K^w - \varepsilon_K^w EP_K^w - \gamma_K^{sw}(Q_K^s/Q_K^w)EQ_K^s - (Q_{Ki}^w/Q_K^w)EQ_{Ki}^w + (Q_{Ke}^w/Q_K^w)EQ_{Ke}^w = EW_K^w,$

(16″) Imported wholesale pork derived demand: $EQ_{Ki}^w - \eta_{Ki}^w EP_{Ki}^w - \tau_K^{rw} EQ_K^w = (Q_{Ki}^w/Q_K^w)EZ_{Ke}^w + EZ_{Ki}^w$,

(17″) Imported wholesale pork derived supply: $EQ_{Ki}^w - \varepsilon_{Ki}^w EP_{Ki}^w = EW_{Ki}^w$,

(18″) Exported wholesale pork derived demand: $EQ_{Ke}^w - \eta_{Ke}^w EP_K^w = EZ_{Ke}^w$,

(19″) Market hog derived demand: $EQ_K^s - \eta_K^s EP_K^s - \tau_K^{ws} EQ_K^w = (Q_{Ke}^w / Q_K^w) EZ_{Ke}^w + EZ_K^s$,

(20″) Market hog primary supply: $EQ_K^s - \varepsilon_K^s EP_K^s = EW_K^s$,

Poultry marketing chain

(21″) Retail poultry primary demand: $EQ_Y^r - \eta_{YB}^r EP_B^r - \eta_{YK}^r EP_K^r - \eta_Y^r EP_Y^r = EZ_Y^r$,

(22″) Retail poultry derived supply: $EQ_Y^r - \varepsilon_Y^r EP_Y^r - \gamma_Y^{wr} EQ_Y^w = EW_Y^r$,

(23″) Wholesale poultry derived demand: $EQ_Y^w - \eta_Y^w EP_Y^w - \tau_Y^{rw} EQ_Y^r = EZ_Y^w$,

(24″) Wholesale poultry primary supply: $EQ_Y^w - \varepsilon_Y^w EP_Y^w + (Q_{Ye}^w / Q_Y^w) EQ_{Ye}^w = EW_Y^w$,

(25″) Exported wholesale poultry derived demand: $EQ_{Ye}^w - \eta_{Ye}^w EP_Y^w = EZ_{Ye}^w$.

Balagtas and Kim (2007) note this model can be expressed in matrix form as **RY = Z**, where **R** is a matrix of model parameters (i.e., elasticities), **Y** is a column vector of endogenous changes in prices and quantities relative to an initial equilibrium, and **Z** is a column vector of percentage changes associated with vaccination protocol adoption. The model defines proportional changes in equilibrium prices and quantities for each evaluated market level and species in response to exogenous changes corresponding to vaccination introduction. These proportional changes are identified as:

(26) $\mathbf{Y} = \mathbf{R}^{-1}\mathbf{Z}$.

We use producer surplus to quantify the net economic impact of vaccination adoption. Changes in producer surplus created by introducing vaccinations can be calculated in terms of changes in prices and quantities identified by the EDM as:

(27) $\Delta PS_a^m = P_a^m Q_a^m (EP_a^m + EW_a^m)(1 + 0.5 EQ_a^m)$.

where producer surplus is denoted by PS (Lusk and Anderson 2004). The superscript m denotes the market level (i.e., r = retail, w = wholesale (processor/packer), s = fed cattle (feeding), and f = feeder (farm level)) and subscript a denotes the industry/species evaluated (i.e., beef, pork, or poultry). Change in total producer surplus is the sum of the change in producer surplus from each market level for a species, $\Delta PS_a = \sum_m \Delta PS_a^m$.

Solutions to equation (26) require elasticity estimates for the matrix of parameters (**R**). Identifying these estimates by econometrically estimating structural supply and demand equations for the 25-equation EDM is problematic. As in most EDM applications, direct estimation of elasticities is prohibited by the large number of equations and by identification problems in jointly estimating supply and demand relationships (Brester et al. 2004). However, given the *E. coli* vaccination results in relatively small aggregate market shifts (in proportional terms), we follow standard EDM procedures and utilize elasticity estimates reported in the published literature.

We simulate our model annually for ten consecutive years to allow for adoption over time by producers of the *E. coli* vaccination technology. Consistent with historical beef cattle cycles, we assume that it takes the marketplace ten years to fully adjust from short-run to long-run relationships. Ten years of market effects were simulated by linearly adjusting all elasticities between short-run (year 1) and long-run (year 10) using elasticity estimates employed by Pendell et al. (2010)[f]. The supply, demand, and quantity transmission elasticities used are equivalent to those used by Schroeder and Tonsor (2011). Similarly, base price and quantity values are necessary to estimate surplus calculations. The market price and quantity values are annual average values for calendar year 2012 as reported by the Livestock Marketing Information Center (LMIC).

Our analysis assumes 10% of fed cattle are vaccinated in year 1, 25% in year 2, 50% in year 3, and 90% in years 4–10. This reflects a typical "S-curve" adoption pattern where adoption increases rapidly upon introduction of the technology with a plateau corresponding to the fact that few technologies are every completely adopted by all parties in a heterogeneous industry.

Competing interests
The authors declare that they have no competing interests.

Acknowledgements
Partial funding assistance is acknowledged from the federal-State Marketing Improvement Programs USDA and the USDA STEC CAP grant.

References

AgManager.Info. 2013. Livestock & Meat Marketing: Meat Demand Tables, Charts, and Videos. Available at: http://www.agmanager.info/livestock/marketing/Beef%20Demand/default.asp

Antle JM (1999) Benefits and costs of food safety regulation. Food Policy 24:605–623

Antle JM (2000) No such thing as a free safe lunch: the cost of food safety regulation in the meat industry. Am J Agric Econ 82:310–322

Bailey D (2007) Political economy of the U.S. Cattle and beef industry: innovation adoption and implications for the future. J Agric Resour Econ 32:403–416

Balagtas JV, Kim S (2007) Measuring the effects of generic dairy advertising in a multi-market equilibrium. Am J Agric Econ 89:932–946

Blasi D, Brester G, Crosby C, Dhuyvetter K, Freeborn J, Pendell D, Schroeder T, Smith G, Stroade J, Tonsor G (2009) Benefit-Cost Analysis of the National Animal Identification System, Final Report submitted to USDA-APHIS on January 14, 2009., Available at: http://www.naiber.org/Publications/NAIBER/BC.analysis.NAIS.pdf (accessed August 5, 2013)

Brester GW, Marsh JM, Atwood JA (2004) Distributional impacts of country-of-origin labeling in the U.S. Meat industry. J Agric Resour Econ 29:206–227

Callaway TR, Edrington TS, Loneragan GH, Carr MA, Nisbet DJ (2013) Review: current and near-market intervention strategies for reducing Shiga toxin-producing *Escherichia coli* (STEC) shedding in cattle. Agric Food Anal Bacteriol 3:103–120

Cull CA, Paddock ZD, Nagaraja TG, Bello NM, Babcock AH, Renter DG (2012) Efficacy of a vaccine and direct-Fed microbial against fecal shedding of *escherichia coli* O157:H7 in a randomized Pen-level field trial of commercial feedlot cattle. Vacccine 30:6210–6215

Dodd CC, Sanderson MW, Jacob ME, Renter DG (2011) Modeling preharvest and harvest interventions for escherichia coli O157 contamination of beef cattle carcasses. J Food Prot 74(9):1422–1433

Ferrier PM, Buzby JC (2013) The economic efficiency of sampling size: the case of beef trim. Risk Anal 33:368–384

Gabbett RJ (2010) "Cargill Settles E. coli Lawsuit with Stephanie Smith." Meatingplace.com. May 13., Available at: http://www.meatingplace.com/Industry/News/Details/16513

Hurd HS, Malladi H (2012) "An outcomes model to evaluate risks and benefits of escherichia coli vaccination in beef cattle." Foodborne Pathog Dis 9(10):952–961

Livestock Marketing Information Center. 2013. Available at: http://www.lmic.info/.

Lueger A, Schroeder TC, Renter DG (2012) "Feedlot Costs of Vaccinating Cattle for *E.coli*." K-State Department of Agricultural Economics. Publication: TCS-December 2012., Available at: http://www.agmanager.info/livestock/budgets/production/beef/TCS_FactSheet_EcoliVaccination_12-07-12.pdf. Accessed on December 2, 2013

Lusk JL, Anderson JD (2004) Effects of country-of-origin labeling on meat producers and consumers. J Agric Resour Econ 29:185–205

Marsh TL, Schroeder TC, Mintert J (2004) "Impacts of meat product recalls on consumer demand in the USA," Appl Econ 36:897–909

Matthews L, Reeve R, Gally DL, Low JC, Woolhouse MEJ, McAteer SP, Locking ME, Chase-Topping ME, Haydon DT, Allison LJ, Hanson MF, Gunn GJ, Reid SWJ (2013) "Predicting the public health benefit of vaccinating cattle against Escherichia coli O157.". Proc Natl Acad Sci U S A 110:16265–162770, Available at: www.pnas.org/cgi/doi/10.1073/pnas.1304978110

Moghadam AK, Schmidt C, Grier K (2013) The impact of *E.Coli* O157:H7 recalls on live cattle futures prices: revisited. Food Policy 42:81–87

Pendell D, Brester G, Schroeder T, Dhuyvetter K, Tonsor GT (2010) Animal identification and tracing in the United States. Am J Agric Econ 92:927–940

Pendell DL, Tonsor GT, Dhuyvetter KC, Brester GW, Schroeder TC (2013) Evolving U.S. Beef Export Market Access Requirements for Age and Source Verification." Food Policy. Available at: http://www.sciencedirect.com/science/article/pii/S030691921300064X#

Perry B, Marsh TL, Jones R, Sanderson MW, Sargeant JM, Griffin DD, Smith RA (2007) "Joint product management strategies on E. Coli O157 and feedlot profits,". Food Policy 32:544–565

Piggott NE, Marsh TL (2004) "Does food safety information impact US meat demand?". Am J Agric Econ 86:154–174

Scallan E, Hoekstra RM, Angulo FJ, Tauxe RV, Widdowson M-A, Roy SL, Jones JL, Griffin PL (2011) Foodborne illness acquired in the United States—major pathogens. Emerg Infect Dis 17:7–15

Scharff RL (2010) Health-Related Costs from Foodborne Illness in the United States." Produce Safety Project. Georgetown University. Available at: http://www.pewhealth.org/reports-analysis/reports/health-related-costs-from-foodborne-illness-in-the-united-states-85899367904

Schroeder TC, Tonsor GT (2011) Economic impacts of zilmax® adoption in cattle feeding. J Agric Resour Econ 36:521–535

Scott C (2012) "Maple Leaf Foods Distributes Listeriosis Settlement Checks." Meatingplace.com. February 7. Available at: http://www.meatingplace.com/Industry/News/Details/30497

Smith GG, Goebel SE, Culbert CR, Guilbault LA (2013a) Reducing the public health risk of escherichia coli O157 exposure by immunization of cattle. Can J Public Health 104:e9–e11, Available at: http://www.ncbi.nlm.nih.gov/pubmed/23618124

Smith BA, Fazil A, Lammerding AM (2013b) A risk assessment model for *escherichia coli* O157:H7 in ground beef and beef cuts in Canada: evaluating the effects of interventions. Food Control 29:364–381

Snedeker KG, Campbell M, Sargeant JM (2012) A systematic review of vaccinations to reduce the shedding of Escherichia coli O157 in the faeces of domestic ruminants. Zoonoses Public Health 59:126–138

Thomson DU, Loneragan GH, Thornton AB, Lechtenberg KF, Emery DA, Burkhardt DT, Nagaraja TG (2009) "Use of a sidephore receptor and porin proteins-based vaccine to control the burden of *escherichia coli* O157:H7 in feedlot cattle." Foodborne Pathog Dis 6(7):871–877

Tonsor GT, Mintert J, Schroeder TC (2010) "U.S. Meat demand: household dynamics and media information impacts." J Agric Resour Econ 35:1–17

U.S. Department of Agriculture, Economic Research Service. 2010. "Foodborne Illness Cost Calculator: STEC O157:H7." Available at: http://www.ers.usda.gov/Data/FoodBorneIllness/ecoli intro.asp

Varela NP, Dick P, Wilson J (2013) Assessing the existing information on the efficacy of bovine vaccination against *Escherichia coli* O157:H7 – a systematic review and meta-analysis. Zoonoses Public Health 60:253–268

Vogstad AR, Moxley RA, Erickson GE, Klopfenstein TJ, Smith DR (2013) Assessment of heterogeneity of efficacy of a three-dose regimen of a type III secreted protein vaccine for reducing STEC O157 in feces of feedlot cattle. Foodborne Pathog Dis 10:678–683

Weise E (2011) "Who Should Pay to Make Ground Beef Safe from E. coli?" In: USA Today, November 28. Available at: http://usatoday30.usatoday.com/money/industries/food/story/2011-12-01/safe-meat/51447546/1

Wohlgenant, Michael K (1993) Distribution of Gains from Research and Promotion in Multi-Stage Production Systems: The Case of the U.S. Beef and Pork Industries. American Journal of Agricultural Economics 75(3):642-51.

Zingg A, Siegrist M (2012) People's willingness to eat meat from animals vaccinated against epidemics. Food Policy 37(3):226-231.

11

Price relations between international rice markets

Adam John

Correspondence: adam.dr.john@
hotmail.com
Institute of Agricultural and Food
Policy Studies, Universiti Putra
Malaysia, Putra Infoport, Jalan
Kajang-Puchong, 43400 UPM, Ser-
dang, Selangor, Malaysia

Abstract

International rice markets are seen as volatile due to the thin nature of the market which is believed to be exacerbated by a low level of substitution between major rice export markets. In other words, this perceived lack of price transmission amongst international rice markets is believed to further thin out an already thin world rice market. The paper tests for price transmission between five major rice exporting markets representing Asia and the Americas over the past decade. It uses a vector autoregressive framework and performs Granger and Toda-Yamamoto causality tests and generalized impulse response functions to interpret the model's results. The findings suggest that price transmission exists across these major rice export markets with price relations being the most widespread between Asian markets. Furthermore, the direction of price transmission suggests that Asian prices act as price leaders for North and South American prices. While it is not clear whether there is a price leader amongst the Asian export markets, Vietnam has the most extensive price relations with other export markets which would suggest that the Vietnamese rice export price is a more suitable world reference price than the Thai export price. An implication of the presence of price relations between rice export markets is that the world rice market is not as fragmented as generally perceived in the literature. However, it can also explain why international rice prices are so sensitive to the volatile trading behavior of major markets.

Keywords: Rice export markets; Price transmission; Causality tests; Impulse response functions; World reference price

Background

The world rice market is seen as a thin market which is used to explain its volatile nature (Siamwalla and Haykin, 1983; Gibson, 1994; Wailes, 2002; Nielsen, 2003; Wailes, 2005; Headey, 2011; and Rapsomanikis, 2011). Two explanations are usually given for why it is a thin market. Firstly, it has been pointed out that the proportion of world rice production traded internationally represents just seven percent of the market (Wailes, 2005; Headey, 2011). In such an event, prices do not reflect the supply and demand conditions of the market (Tomek and Robinson, 1990), it increases search costs and can lead to excessive price volatility (Anderson et al. 2007). The second argument given is that international rice markets have a low level of substitution (Petzel and Monke, 1980; Siamwalla and Haykin, 1983 Rastegari-Henneberry, 1985; Cramer et al. 1993; Jayne, 1993; Chan, 1997, Wailes, 2005; Dawe, 2008) which fragments the world market into even smaller unrelated markets and makes it harder to discover price information (Jayne, 1993).

Due to the latter argument, one would expect price relations between international rice markets to be weak with little price transmission taking place across different geographical rice markets. The behavior of international rice prices and trade, however, would suggest that rice markets must be related to some extent, at least in contemporary times. Firstly, the huge surge in international rice prices in 2007-2008 was felt worldwide which transmitted over to most Asian, African and Latin American domestic markets (Demeke et al. 2011). Secondly, major rice exporting nations are competing in some of the same markets, particularly in Africa and parts of Latin America[a].

The aim of this study is to test to what extent international rice markets are related to each other by measuring price transmission using the monthly export rice prices of five major markets taken from FAO's Global Information and Early Warning System database. The prices are specified into a vector auto regression (VAR) model and the results are interpreted using Granger and Toda-Yamamoto causality tests as well as generalized impulse response functions (IRF). The causality tests are used to verify whether price transmission exists between several major rice export markets from Asia and the Americas. Furthermore, generalized impulse response functions are simulated to provide information on the magnitude and persistence of the price transmission which takes place between international rice markets. Understanding the extent of price relations between international rice markets gives an indication of how well the world rice market functions and even whether it is appropriate to view international rice markets as a single market. It also gives an idea of how competitive exporting countries are and since the causality tests can indicate the direction of price transmission, the testing procedure can also provide an insight into whether there are certain markets which act as price leaders.

Rice exporting nations are often reported to distort their export prices in order to stabilize their own domestic markets (Dorosh, 2009; Gilbert and Morgan, 2010; Timmer, 2010; Dawe and Slayton, 2011; Demeke, et al. 2011; Headey, 2011). Headey (2011) and Gilbert (2013) both believe such behavior played a major role in the huge price hikes seen in international rice prices and more recently, the perceived unreliability of international rice markets has led to the promotion of rice self sufficiency polices in many traditional rice importing countries under the guise of national food security (Stage and Rekve, 1998; Chand, 2006; Xiufang and Dwyer, 2008; Seck et al. 2010; Demeke et al. 2011). A problem with these policies is that they are expected to make the world rice market even thinner and therefore more volatile (Jayne (1993; Wailes, 2005). If export prices are found to be related then it would suggest that exporting nations are to some extent responsive to the market's price movements and therefore not as price distortive as perceived.

The study not only tests for the existence of price transmission between international rice markets but also the direction of price transmission. The direction is also of interest as this information can provide further insights such as whether there exists a suitable reference price for the world rice price. International organizations such as the International Monetary Fund (IMF) tend to use the Thai export FOB price as a proxy for the world rice price which is sensible since Thailand is traditionally the largest rice exporter. The estimation procedure of this study also allows for testing whether the Thai price is a suitable proxy or whether international rice prices are too isolated and fragmented for there to be a credible world rice price.

Literature review

There are numerous varieties of rice which are consumed and traded; however, conventional indica varieties make up 85 per cent of world rice consumption and 80 per cent of world rice trade (Jayne, 1993; Dawe, 2008). The general view has been that prices of the different rice varieties, such as indica and japonica markets are unrelated (Petzel and Monke, 1980; Siamwalla and Haykin, 1983 Rastegari-Henneberry, 1985; Dawe, 2008). However, Falcon and Monke (1980) expected indica prices between Asian and American export markets to be integrated and Dawe (2008) also expected these markets were strongly related throughout the 1980s. In more contemporary times however, Dawe (2008) believes that American and Asian indica rice markets act more like two separate commodity markets which he argues can be explained by the fact that the US does not export rice to Asian markets.

Testing for price transmission under a market integration framework using cointegration tests has been a common practice in the literature. However, most studies related to rice markets have focused on testing for price transmission between international and domestic rice markets in order to assess how integrated domestic rice markets are with international markets. However, Ghoshray (2008) used this approach to test for price relations between two major rice export markets and found evidence which suggested that Thailand and Vietnam's export prices were asymmetrically cointegrated. That is to say that although their prices share a long-run relationship, price transmission is asymmetric, which is a sign of uncompetitive behavior. He found the prices to be asymmetric in the sense that price adjustments occur faster when the price differentials are in decline rather than when they are increasing. In addition, Ghoshray (2008) looked for whether there was a rice price leader between Thailand and Vietnam. He concluded that Thailand acts as a price leader for higher quality grades but also responds to Vietnamese price movements to some extent in the short-run.

Using causality tests and IRFs to interpret the results of a VAR model for assessing price transmission between Thailand's domestic and export rice markets, John (2013) found price transmission to be bi-directional. However, the magnitude and persistence of price transmission based on the IRF results between the two markets was found to vary substantially, suggesting that although Thailand's paddy pledging program distorts prices and hinders price transmission between Thai domestic and export prices in the short-run to some extent, Thai export price movements eventually transmit through to domestic prices substantially.

Methods

Research focused on measuring price transmission between spatial markets has most often been concerned with testing for market integration which uses the Law of One Price (LOP) as a theoretical framework. LOP asserts that at all points of time the relationship between two markets is as follows:

$$p1t = c + p2t \qquad (1)$$

where due to instantaneous competitive arbitrage the price differential between market p_1 and market p_2 at time t is the transfer cost (c) of the product to each of the markets[b]. The notion stated in equation (1) is a strong form of market integration since markets deviating from this are considered not to be integrated and therefore no price transmission takes place. If this relationship holds then it is said that markets are integrated and full transmission takes place. A weaker form of LOP allows the price relationship to be as follows:

$$|p2t-p1t| \leq c \qquad (2)$$

where the price differential between markets may be less than the transfer cost but efficient arbitrage will not allow the deviation to exceed the transfer cost. Equation (2) is less restrictive in the sense that transmission can take place even when the price differential is less than the transfer cost and suggests an equilibrium condition.

$$|p2t-p1t| = c \qquad (3)$$

One might expect the condition in equation (3) to occur in the long run since traders would leave the market if price deviations did not cover the transfer cost. However, the condition in equation (2) can occur at least in the short run. This is essentially relaxing the assumption of instantaneous arbitrage. In terms of price transmission, this may mean that while transmission may not take place fully in the short run between integrated markets, full transmission would be expected to occur in the long run. In the context of price transmission, cointegration can be seen as the empirical equivalent of the theoretical concept of a weaker form of LOP, where it is accepted that while price transmission does not adjust instantly, in the long run the full extent of the transmission takes place. However, the traditional linear cointegration tests have been criticized because they do not consider the existence of non-stationary transfer costs. Ghoshray (2008) overcomes this criticism by using threshold autoregressive (TAR) models which allow for asymmetric error corrections in the cointegration tests.

This study adopts less restrictive estimation techniques for measuring price transmission, namely causality tests and impulse response functions. While testing for cointegration is a common procedure, the more conventional cointegration methods are rather restrictive as they require all of the price series to be I(1) non-stationary processes. Cointegration techniques have also been criticized as not being suitable tools for measuring market integration on several grounds including the fact that they do not consider the trade behavior of countries (Barrett and Li, 2002). One of the largest concerns with the LOP framework is that transfer costs between markets can be so great that it does not seem appropriate to test for market integration but instead simply test to see whether markets have in fact any price relationship. Transportation is perhaps the most obvious transfer cost which affects price transmission; however the factor which has been given the most attention in the literature, when assessing spatial markets, is border and domestic policies (Conforti, 2004). Conforti (2004) sees product homogeneity, that is to say, the level of substitutability between products across different markets, to be particularly important to world rice markets. If rice markets have a high level of differentiation, prices across markets are likely to differ substantially. Other key factors considered important to spatial price transmission are: transaction costs, exchange rates, market power, and increasing returns to scale in production (Conforti, 2004; Ghoshray, 2011; Gilbert, 2011). In this study, the aim is not to test for market integration but merely provide some insight into the competiveness of major rice export markets by testing whether their price movements are related.

Vector autoregression (VAR)

A VAR is a system of equations which is rooted in the Box-Jenkins procedure. It presumes that the best way of predicting movements in x are the past values of x as well as the past values of other variables. Since no restrictions are made in the system, everything is assumed to affect everything and there are no endogeneity issues since there are no contemporaneous explanatory variables. Its simplicity makes it an attractive model as compared to the traditional structural equation models which demand sound economic theory surrounding the relationship between the studied variables. The absence of a solid theoretical background linking the variables within the VAR system has also been regarded as a critical weakness of the model which has led to the popularity of restricted VAR models. For this study, it is felt that there should be no restrictions on the relations between any of the export prices as the assumption that all of the prices affect one another is sensible if the markets are competitive. Therefore, specifying a restricted VAR would be inappropriate in this case.

The equations of the VAR are estimated using the ordinary least squares (OLS) procedure therefore all price series within the system must be stationary processes otherwise the OLS assumptions are violated. Knowledge about the stationarity of the price series is therefore required. If the prices are stationary then the price series can enter the VAR system in their level form. However, if any of the prices are non-stationary, the series will need to be first-differenced before they can enter the VAR system in order to ensure the OLS assumption of stationarity is not violated. As an example, a simple 2×2 VAR can be specified as follows:

$$y_t = a_1 + \sum_{i=1}^{k} \phi_{1i} y_{t-i} + \sum_{i=1}^{k} \theta_{1i} x_{t-i} + u_{1t} \tag{4}$$

$$x_t = a_2 + \sum_{i=1}^{k} \phi_{2i} y_{t-i} + \sum_{i=1}^{k} \theta_{2i} x_{t-i} + u_{2t} \tag{5}$$

where equation (4) specifies the export price of country y as a function of its own past price values up to k lags, the past values of the export price of country x up to k lags and a white noise error term (u_1). Equation (5) specifies the same explanatory variables for country x.

Causality tests

In the Granger sense, causality is the ability of past values of x predicting the contemporaneous movements in y. If this holds, x can be said to Granger-cause y. Within the field of price transmission, this can be seen as the ability of the past price movements of one market predicting the contemporaneous movements of another price. The Granger causality test is basically an autoregressive specification of y where the past values of another variable, x, are added to the autoregressive equation. In other words, it tests whether the removal of the lags of x from the autoregressive equation leads to a loss of information which explains the current movements in y. The Granger test is essentially a Wald test which restricts all of the lags of the explanatory variable we are interested in to zero within the autoregressive equation. For equations 4 and 5 the null and alternative hypotheses (H_0 and H_a) are as follows:

$$H_{10} : \theta_{11} = \theta_{12} = \cdots = \theta_{1k} = 0; H_{1a} : \text{At least one } \theta_1 \text{ is not zero}$$

$$H_{20} : \phi_{21} = \phi_{22} = \cdots = \phi_{1k} = 0; H_{2a} : \text{At least one } \phi_2 \text{ is not zero}$$

In this example, one could infer that a rejection of H_{10} and H_{20} suggests there is bi-directional transmission between the export prices while failing to reject H_{10} and H_{20}

would suggest there is no price transmission. Meanwhile, rejecting H_{10} and not H_{20} would suggest that prices only transmit unidirectionally from export price x to export price y, while rejecting H_{20} and not H_{10} would suggest the opposite.

If the Granger causality tests are performed on the first-differences of the price series then the tests provide information on the relationship between the price changes and can be seen as short run price transmission since the long run information is taken out of the price series when they are first-differenced. There is, however, another causality test which can be performed even if the system of equations includes non-stationary processes. The Toda-Yamamoto causality test is similar to the Granger test but has the advantage of allowing non-stationary series to be included in the test procedure therefore the series do not need to be first-differenced. This means that the series contain their long-run information so the Toda-Yamamoto test can provide insight into price transmission which is not restricted to the short-run. The only preliminary information needed before performing the Toda-Yamamoto causality test is the maximum order of integration of the variables (d_m) included in the VAR system. Once the optimal number of lags (k) is selected for the model, the VAR is specified as a $VAR(k + d_m)$. For instance, if the optimal number of lags is three and the maximum order of integration is one, the model is specified as a $VAR(3 + 1)$.

The Toda-Yamamoto causality test then follows the same procedure as the Granger causality test where a Wald test is performed to restrict all of the lags of the selected explanatory variable up to k to zero. Equation (4) would therefore be modified as follows:

$$y_t = a_3 + \sum_{i=1}^{k+dm} \phi_{3i} y_{t-i} + \sum_{i=1}^{k+dm} \theta_{3i} x_{t-i} + u_{3t} \tag{6}$$

The hypotheses of the Toda-Yamamoto test are as follows:

$$H_{30} : \theta_{31} = \theta_{32} = \cdots = \theta_{3k} = 0; H_{3a} : \text{At least one } \theta_3 \text{ is not zero}$$

whereby the coefficients included in the test go up to k and not d_m. The results can then be interpreted in the same way as the Granger causality results.

Impulse response functions

A weakness of the causality tests is that the values of the coefficients are difficult to interpret within the VAR system therefore these tests cannot tell us anything about the magnitude of the price transmission, only whether price transmission exists or not. Impulse response functions (IRF) are commonly used to assess what impact each of the variables included within the VAR system have on one another. It does this by assessing the error terms of the equations which it sees as shocks. IRFs simulate the effect a shock originating in one variable has on the other variables in the VAR system and is able to quantify its impact over proximate time periods. Cholesky factorization is used to identify where the shocks are coming from which is done using the recursive structure of the values of the variance-covariance matrix elements to select the restrictions and therefore identify the origin of the shocks. In other words, the assumption is made that the shocks from the reduced form equations simulate the structural shocks of the VAR system.

A major issue with using the recursive structure for identifying the shocks is that the IRFs become sensitive to the ordering of the variables within the VAR system if the error terms are highly correlated. One way to get over this problem is by ordering the variables from the most exogenous to most endogenous variable within the system. However this requires in depth knowledge of the relationship between the variables. In the case of this study, it does not seem appropriate to leave it up to the researcher's *a priori* judgement on the exogeneity of the export prices as this is precisely what the study is attempting to test. A solution is to use the Generalised IRFs as these are not sensitive to the ordering of the variables, and are therefore used in this study. IRFs can simulate how one variable responds to a one standard deviation shock in its own stochastic process or any other variable within the VAR system over a designated number of time periods after the initial shock. IRFs can therefore provide information on the magnitude of price transmission as well as its persistence and direction. For this study, generalized IRFs will be simulated in the cases where export prices are found to Granger cause other export prices as they can provide further information to supplement the causality tests as to the extent of price transmission between international rice markets.

Preliminary analysis of data

The export prices of five major rice markets are used in this study which were taken from the Food and Agricultural Organisation's Global Information Early Warning System (GIEWS) database. The prices are measured in a common currency, namely monthly FOB export prices in nominal US dollars. The Thai and Vietnamese prices are the five per cent broken rice prices, the American price is the two to four per cent broken price, the Argentinean price is the ten per cent broken price, and the Pakistani price is the twenty five per cent broken price. The prices are therefore seen as representing the higher quality indica grades of rice apart from the Pakistani price. While the higher quality Pakistani price was not available, it is assumed that the higher quality Pakistan price closely follows price movements in the lower quality Pakistani price since this is the case for Thai and Vietnamese rice export prices. The Uruguayan export price is probably a more important rice price for South American markets than the Argentinean price, however, data was only available from 2006 which would have severely cut the available number of observations included in the study therefore the study assumes that Argentinean prices are related to Uruguayan price movements and are therefore suitable for the analysis. Indian export prices are also available; however, due to the Indian government's policy of temporal rice export bans there are many breaks in the price series therefore Indian prices cannot be included in the study.

Three unit root tests are used to provide insight into the stationarity of the export prices. The Augmented Dickey-Fuller (ADF) and Phillips-Perron (PP) tests are seen as the conventional unit root tests whereas the Zivot-Andrews (ZA) test is also included as this test considers the existence of a structural break in the time series. An advantage of the ZA test is that it endogenously determines the date of the structural break rather than leaving it up to the researcher's judgement to decide on the date. From observing Figure 1, the behavior of all five price movements suggests that there may have been a structural break in the time period 2007-2008. *A priori* knowledge of world rice prices would suggest this to be true as this period is often referred to as the

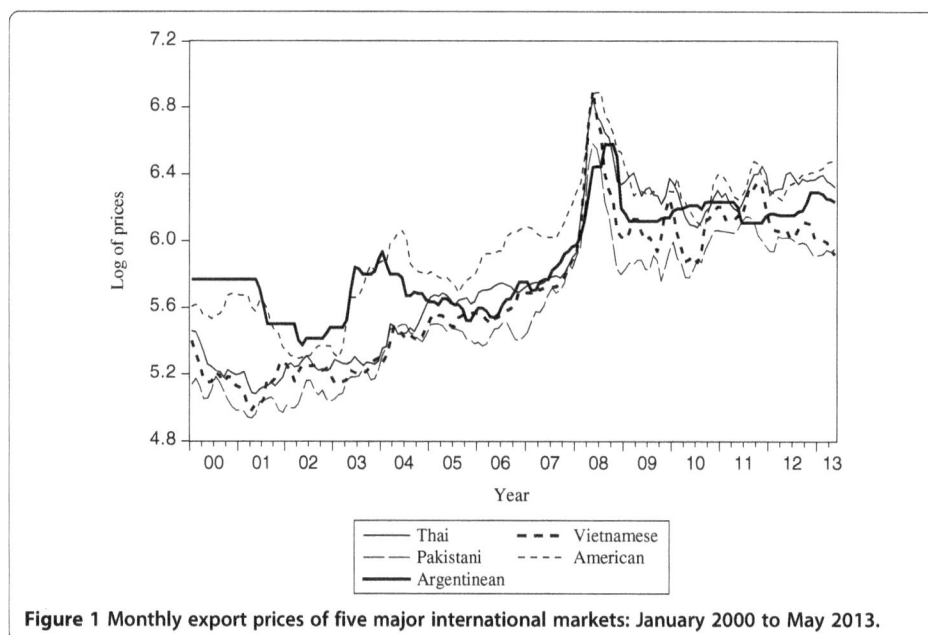

Figure 1 Monthly export prices of five major international markets: January 2000 to May 2013.

'food crisis' when international food commodity prices including rice increased to unprecedented levels.

According to the result of Table 1, the ADF give mixed results whereby evidence is found to suggest that the Asian rice prices are stationary while the American and Argentinean prices are I(1) non-stationary processes. On the other hand, the PP tests find all of the prices to be I(1) non-stationary processes. The ZA results, however, conform to the ADF results which find the three Asian prices to be stationary processes when considering the existence of a structural break while the American and Argentinean prices are non-stationary. Since the power of the conventional unit root test is lowered in the presence of a structural break (Perron, 1989) the study considers the ZA results to be the most suitable to follow. Furthermore, since both the ADF and PP results did not find any of the prices to have a higher order of integration than one, one is considered to be the maximum order of integration for the Toda-Yamamoto causality procedure.

Results

Two VAR models were specified. The first used the first-differences of the five export prices while the second contained the prices in their level form. Two VARs were

Table 1 Unit root tests[a]

Export Price	Level				First-difference	
	ADF	PP	ZA	Structural break	ADF	PP
Thailand 5%	−3.380*	−2.884	−6.552***	2008 m02	−7.556***	−7.264***
Vietnam 5%	−3.804**	−2.796	−6.017***	2008 m02	−7.251***	−5.923***
Pakistan 25%	−3.794**	−2.813	−5.692***	2007 m09	−7.123***	−7.096***
U.S 2-4%	−3.102	−2.589	−4.141	2007 m09	−6.984***	−7.025***
Argentina 10%	−2.808	−2.422	−4.001	2007 m07	−7.818***	−7.707***

[a] The time period of the price series is from January 2000 to May 2013. The prices are in their logarithmic transformations. A trend and intercept were included in the unit root tests. *, **, ***, signify a ten, five, and one percent level of significance of the unit root tests.

needed so Granger causality tests could be made using the first VAR system and could provide information on the short run price transmission between international rice markets since this system uses the price changes of the export prices. Meanwhile, the second VAR system was used for the Toda-Yamamoto causality procedure which can provide insight into price transmission which was not restricted to the short term.

The Granger causality results are in Table 2. They suggest that short run price transmission exists across the Asian markets with bidirectional price transmission between Thailand and Vietnam as well as between Pakistan and Vietnam. Price transmission occurs in the short run from Pakistan to Thailand; however, it is not bidirectional. The Granger test did not find evidence to suggest that Thai prices transmitted to Pakistani prices in the short run.

Surprisingly, no short run transmission was found between the US and any of the Asian markets which suggests that the former is completely unrelated to the latter markets even though their rice markets are of a similar grade and they compete in some of the same markets in Africa and parts of Latin America. The only market American price movements responded to in the short run was Argentina; however the price transmission was not bidirectional. On the other hand, the Argentinean market was found to be related to two of the Asian markets, namely Vietnam and Pakistan. However, Argentinean prices did not transmit over to any of the Asian markets, possibly because the former is dwarfed by the latter markets.

Overall, the most price related markets were Vietnam and Pakistan as they both transmitted their price movements to all of the other markets apart from the US. In addition, Vietnam was the only market to respond to the price movements of all of the other markets except for the US which would suggest that Vietnam is the most suitable reference price for international rice prices amongst the five prices, in the short run at least.

According to the Toda-Yamamoto causality results, price transmission which considers both the short and long-run price movements is similar amongst the Asian markets vis-à-vis short run price transmission which was tested using the Granger procedure. What is interesting is that even considering the long run information in the

Table 2 Granger causality tests[b]

Export prices	→ Thailand	→ Vietnam	→ Pakistan	→ U.S.	→ Argentina
Thailand →	x	20.494***	4.265	3.086	1.776
		(0.001)	(0.385)	(0.687)	(0.879)
Vietnam →	17.269***	x	19.656***	8.238	14.770**
	(0.004)		(0.001)	(0.144)	(0.012)
Pakistan →	12.952**	10.057*	x	7.189	11.127**
	(0.024)	(0.074)		(0.207)	(0.049)
U.S. →	1.248	1.483	2.108	x	4.382
	(0.940)	(0.915)	(0.834)		(0.496)
Argentina →	4.013	3.118	5.809	14.838**	x
	(0.548)	(0.682)	(0.325)	(0.011)	

[b] The figures in the boxes are the Wald test statistics and the p-values are in brackets. *, **, ***, signify a ten, five, and one percent level of significance of the Granger causality tests. The arrows indicate the direction of price transmission. According to the Schwarz Information Criterion, the optimal number of lags is one, however five lags are required in order to ensure the error terms are not serially correlated therefore a VAR(5) model is specified. There are 156 observations.

price series, the Toda-Yamamoto results suggest that Thai prices do not transmit over to Pakistani prices. Having said that, overall, the results of both causality procedures suggest that the Asian export markets are affected and respond to each others' price movements.

American prices do not transmit over to any of the other export prices even when considering the long run information of the price series, however, American prices do respond to the Thai and Vietnamese prices as well as Argentinean prices according to the Toda-Yamamoto results. This suggests that Asian and US rice prices have been related over the past decade. However, the Toda-Yamamoto and Granger test results suggest that Asian markets do not respond to North and South American rice prices. Asian rice prices may therefore be seen as leading the price movements of international rice prices for other regions which is understandable since Asian rice exports make up the bulk of world rice trade.

In line with the Granger results, the Toda-Yamamoto tests suggest that Vietnam has the most far reaching effect on other international rice prices as the results in Table 3 suggest that Vietnamese prices transmit to all four of the other prices included in the analysis. On the other hand, Thai prices only transmit over to Vietnamese and American prices. The causality results therefore suggest that the Vietnamese price is the most suitable reference price for international rice prices for the higher quality indica market, at least.

Generalized impulse response functions were simulated to interpret the magnitude of price transmission in the cases where price transmission was identified by the Granger causality tests in the first specified VAR system. Figure 2 illustrates the impact of shocks transmitting over to other export markets. The Granger causality tests identified that bidirectional price transmission occurred between Vietnam and Thailand, and Vietnam and Pakistan. The IRFs suggest that the magnitude of the price transmission is fairly symmetrical. However, shocks originating in the Vietnamese prices which transmit over to the Thai and Pakistani prices persist slightly longer than shocks originating in the latter two countries and transmitting to the Vietnamese market. Even though the shocks represent the short run effects since it was the price changes which were used to simulate the IRFs, the magnitude of Vietnamese shocks are statistically

Table 3 Toda-Yamamoto causality tests[d]

Export prices	→ Thailand	→ Vietnam	→ Pakistan	→ U.S.	→ Argentina
Thailand →	x	28.277***	9.522	11.105*	2.322
		(<0.001)	(0.146)	(0.085)	(0.888)
Vietnam →	18.652***	x	18.630***	12.127*	15.591**
	(0.005)		(0.005)	(0.059)	(0.016)
Pakistan →	13.818**	17.082***	x	9.047	9.784
	(0.032)	(0.009)		(0.171)	(0.134)
U.S. →	1.470	0.626	2.184	x	3.555
	(0.962)	(0.996)	(0.902)		(0.737)
Argentina →	3.932	0.920	5.027	15.572**	x
	(0.686)	(0.989)	(0.540)	(0.016)	

[d] According to the Schwarz Information Criterion, the optimal number of lags is two, however six lags are required in order to ensure the error terms are not serially correlated therefore a VAR(6 + 1) model is specified. There are 154 observations. *, **, ***, signify a ten, five, and one percent level of significance for the Toda-Yamamoto causality tests.

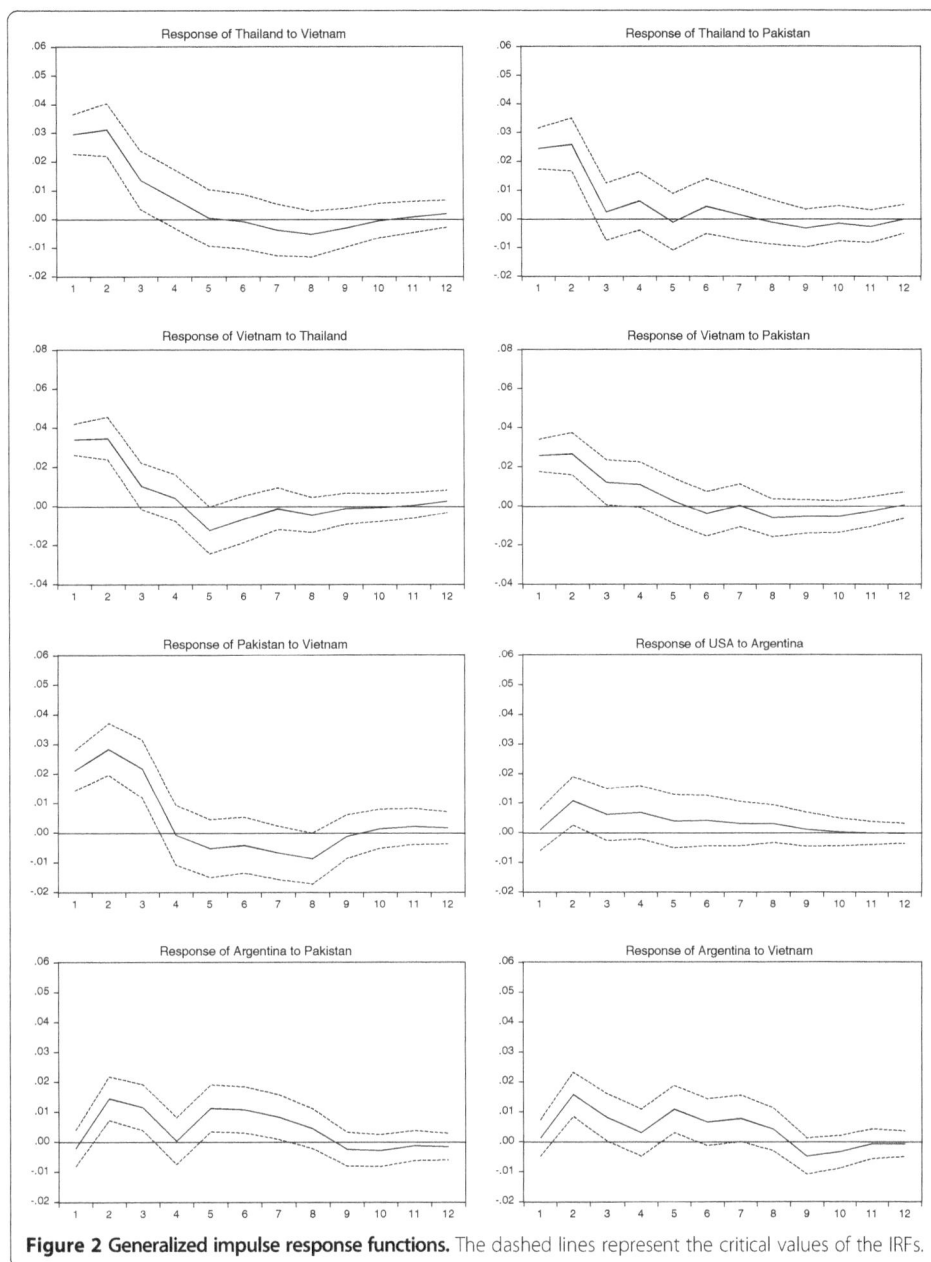

Figure 2 Generalized impulse response functions. The dashed lines represent the critical values of the IRFs.

significant for three months in the other two Asian markets whereas Thai and Pakistani shocks in the Vietnamese market are statistically significant for two months.

Pakistani shocks in the Thai market are roughly as significant in magnitude as Vietnamese shocks, however they only persist for two months, statistically speaking. Although the causality tests found Argentinean prices to transmit over to US prices, the IRF graph shows that its magnitude is barely significant for the second month after the original shock. The last two IRF graphs illustrate the impact of Vietnamese and Pakistani prices on the Argentinean price. What they show is that while the magnitude of this transmission is less than a third of the peak of price transmission between the Asian markets, it is more persistent. For instance, the magnitude of Pakistani and Vietnamese shocks are statistically significant for the second, third, fifth and sixth months and second and fifth months after the original shock in the Argentinean market, respectively.

Perhaps the main finding from the IRFs is that while price transmission occurs between rice markets in Asia and the Americas, the price transmission between Asian markets is far greater than price transmission across the continents. This can be explained by the fact that competition between Asian markets is stronger as they are competing in most of the same markets.

Discussion and conclusions

The paper set out to examine the extent of price relations between major rice exporting nations using causality tests and IRFs to test for price transmission. It is generally perceived in the literature that international rice markets have a low level of substitution which fragments the market and thins out an already thin world rice market. One would therefore expect the prices of different rice export markets to be unrelated.

Both causality tests and the IRFs suggest that Asian rice export markets are related to one another. One would therefore not infer that there is a low level of substitution between these markets which conforms to Ghoshray's (2008) findings that Thai and Vietnamese rice export prices are integrated, albeit asymmetrically. In fact, this study's results would suggest that they are to some extent competitive which one should expect since they compete against each other in many of the same markets.

Although the Granger causality tests reported that Asian and US rice prices are not related in the short term, US prices have, at least over the last decade, responded to price movements of major Asian export markets, according to the Toda-Yamamoto causality tests which were not restricted to using the short run price information. The same results were found for Argentinean rice prices which suggest that the South American rice prices are related to Asian rice prices. While these findings may conform to Dawe's (2008) conclusion, in the short run, that US and Asian rice prices in contemporary times are unrelated, the fact that the Toda-Yamamoto causality results found Asian prices to transmit over to US prices raises some doubt over the view that US and Asian rice markets function as two separate commodity markets. Dawe (2008) is correct that the US does not export indica rice to Asian markets but the US does compete with Asian exporters in some African and Latin American markets, which suggests that there should be some price relations between American and Asian rice export markets.

What the causality tests show is that Asian rice markets are clearly acting as price leaders for the other export markets since the direction of price transmission is from the former and to the latter. This is understandable since the Asian markets are much larger rice producers and dominate international rice trade. The results also show that Vietnamese prices affect the price movements of all of the other export markets, which is not true for Thai prices. This is not necessary contrary to what Ghoshray (2008) reported, namely that between Thailand and Vietnam, the former acts as a price leader. This study's results show that price transmission between these two markets is bidirectional and that the impact according to the IRFs is almost symmetrical. I therefore find no clear price leader between the two markets. However, since the Vietnamese price exhibits price relations between all of the other rice export markets included in this study, which cannot be said for the Thai price, it seems sensible to conclude that the Vietnamese price is a more suitable reference price for international rice prices than the Thai price.

The results of the study imply that domestic and border policies in major rice exporting countries do not inhibit price transmission between many of the rice export markets covered in this study. It is often argued that rice exporting nations distort their export price movements and therefore their trade behavior in order to isolate their domestic rice prices from world price fluctuations. In the case of Thailand, John (2013) reported that while the rice pledging program affects price transmission between Thailand's domestic and export rice markets, price transmission takes place bidirectionally. However, the impact of price transmission from the export market to the domestic market is reduced in the short run but then remains rather persistent.

The impact of border and domestic policy on price transmission between markets depends on the type of policy tool used. Intervention mechanisms such as floor prices could lead to domestic prices being completely unrelated to world prices or being related in a non linear way (Rapsomanikis, Hallam, and Conforti, 2003). Price transmission will only take place if the world price is above the domestic floor price. The floor price will act as a threshold therefore the existence of a floor price may cause price transmission to be asymmetric. Despite the use of minimum rice export prices in Thailand and Vietnam, which act as floor prices, the study's results found these two markets to be related to each other as well as with other markets. This was despite the fact that the methods used in this study did not consider the presence of asymmetric price relationships. This can be explained by the fact that Thai and Vietnamese authorities regularly adjust their minimum export prices in order to keep their rice exports competitive in world markets when they want to release rice stocks. This is precisely what has happened in Thailand in 2013 as the government has continued to lower their minimum export prices in order to make their rice export prices more competitive. While previously, Thailand purposely kept its minimum export prices above world prices, as the cost of the pledging program has built up and surpassed the $16 billion allocated to it, the government has been trying to release the rice stocks it accrued from the rice farmers who participated in the pledging program during 2011 and 2012 in order to recoup part of the cost of the rice pledging program which has cost $8 billion excluding administrative costs (USDA, 2013).

As previously mentioned, like Thailand, Vietnam uses minimum export prices. However, it does not subsidize the price the farmer receives for their rice which is what the pledging program essentially does. The absence of a program similar to the pledging program in Vietnam may explain why Vietnamese rice export prices are more related to other export rice markets than Thai prices according to the study's results. This is because such a policy would require the government to distort trade volumes through holding on to vast sums of rice in public stock holds.

It is unsurprising that Pakistan responds to price movements in other Asian rice markets as it liberalized its rice industry in the later 1990s. US price movements do not transmit over to the other four export markets. This seems to be a rather puzzling finding at first sight since although it is by no means the largest rice exporter; it is still one of the top rice exporters in the world. This may be due to the fact that a substantial amount of rice the US exports is not indica rice but japonica rice. Despite the lack of price transmission from US rice prices to other export markets, US prices do react to price movements in the other major rice export markets according to the Toda-Yamamoto results. There is little reason to doubt these results since the US does not distort its export prices. However, the US heavily subsidizes its rice farmers which push world rice prices down by four to six percent (Griswold, 2006). The US would export far less if not any rice if it was not for it subsidizing its rice farmers.

India was not included in the study; however it is important to mention the impact of Indian domestic policy on the price relations of its rice export market with other international rice markets. Despite being the top rice exporter in recent years, India's trade patterns in rice as well as many other food commodities can be erratic as international markets are seen as a way of offloading surplus stock rather than an opportunity for earning foreign exchange. India provides its farmers with support prices so Indian farmers make their business decisions based on these prices rather than international prices. Minimum export prices are used by India and the government also at times uses export bans in order to secure rice supplies for domestic consumption. Such policies, of course, completely block any price transmission from other export prices to India's prices. It seems intuitive; however, that in periods like what India has experienced recently when it has surplus rice supplies and decides to participate in world rice trade, price transmission occurs as the government drops the minimum export price so their rice prices are competitive with international prices. India's interventionist policies which impede price transmission are likely to remain especially with the implementation of the National Food Security Act passed in parliament in 2013 and involves allocating rice and other grain rations to over 800 million of its inhabitants.

The world rice market may be a thin market but the study's results would suggest that it is not as fragmented as some of the literature proposes, at least when one considers indica rice markets which make up the bulk of world rice production and trade. This is especially the case for Asian rice export markets. An implication of this is that large Asian rice exporting nations have the ability to destabilize the international rice market. In one sense, the study's results demonstrate that governments of major rice export markets do not generally distort their export prices to the extent that these prices remain isolated from other export rice prices. On the other hand, due to the fact that it is such as thin market, the price distortions of one major rice exporting country has the ability to destabilize the international rice market as a whole due to the existence of price transmission between these markets. The unprecedented rise in international rice prices across the world which peaked in 2008 is a striking example which has been argued that the actions of India banning non-basmati rice exports was one of the main causes which initiated the pricing frenzy (Headey, 2011).

To address the concerns of many low income rice import dependent countries, a global rice reserve has been proposed to help stabilize international rice prices during volatile periods (Gilbert, 2013). Since this study would suggest that international rice markets do react to each others' price movements, particularly in Asia, it may seen appropriate that any negotiations of forming such a global rice reserve should involve the two largest rice producers and stockholders, China and India, whereby these countries could play a major role in stabilizing thin world rice markets rather than being a source of volatility.

Endnotes

[a] From observations of annual rice trade data from the UN COMTRADE database.

[b] If $c = 0$ then the prices across the markets would be identical once the prices had been converted into a common currency (Rapsomanikis, Hallam, and Conforti, 2003).

Competing interests
The author declare that he have no competing interests.

Acknowledgements
I appreciate the reviews and comments I received from the two anonymous reviewers.

References
Anderson JD, Hudson D, Harri A, Turner S (2007) A new taxonomy of thin markets. Paper presented at the annual meeting of the Southern Agricultural Economics Association, Mobile, Alabama. 4-7 February 2007
Barrett CB, Li JR (2002) Distinguishing between equilibrium and integration in spatial market analysis. Am J Agric Econ 84(2):292–307
Chan LY (1997) Price instability in the international rice market: its impact on production and farm prices. Dev Pol Rev 15:251–276
Chand R (2006) International trade, food security and the response to the WTO in South Asian countries: discussion paper No. 2006/124. UN University-World Institute for Development Economics Research (UNU-WIDER), Helsinki, Finland
Cramer G, Wailes EJ, Shui S (1993) Impacts of liberalizing trade in the world rice market. Am J Agric Econ 75:219–226
Conforti P (2004) Price transmission in selected agricultural markets: FAO Commodity and Trade Policy Research working paper No. 7. Basic Foodstuffs Service, Commodities and Trade Division, Food and Agricultural Organization, Rome
Dawe D (2008) Have recent increases in international cereal prices been transmitted to domestic economies? The experience of seven large Asian countries. ESA working paper No. 08-03, Agricultural Development Economics Division, The Food and Agriculture Organization of the United Nations, Rome
Dawe D, Slayton T (2011) The world rice market in 2007-08. In: Prakash A (ed) Safeguarding food security in volatile global markets. Food and Agriculture Organization of the United Nations, Rome, pp 171–182
Demeke M, Pangrazio G, Maetz M (2011) Country responses to turmoil in global food markets. In: Prakash A (ed) Safeguarding food security in volatile global markets. Food and Agriculture Organization of the United Nations, Rome, pp 183–210
Dorosh PA (2009) Price stabilization, international trade and national cereal stocks: world price shocks and policy response in South Asia. Food Secur 1:137–149
Falcon W, Monke E (1980) International trade in rice. Food Res Inst Stud 17(3):279–306
Gibson J (1994) Rice self-sufficiency in Papua New Guinea. Rev Mark Agric Econ 62(1):63–77
Gilbert CL (2011) Grains price pass-through, 2005-09. In: Prakash A (ed) Safeguarding food security in volatile global markets. Food and Agriculture Organization of the United Nations, Rome, pp 127–148
Gilbert CL (2013) International agreements to manage food price volatility. Global Food Secur 1:134–142
Gilbert CL, Morgan CW (2010) Food price volatility. Philos Trans R Soc 365(1554):3023–3034
Ghoshray A (2011) Underlying trends and price transmission of agricultural commodities: ADB Economics working paper series No. 257. Asian Development Bank, Metro Manila, Philippines
Ghoshray A (2008) Asymmetric adjustment of rice export prices: the case of Thailand and Vietnam. Int J Appl Econ 5(2):80–91
Griswold D (2006) Grain drain: the hidden cost of U.S. rice subsidies No. 25. Center for Trade Policy Studies, Cato Institute, Washington, DC
Headey D (2011) Rethinking the global food crisis: the role of trade shocks. Food Policy 36:136–146
Jayne TS (1993) Sources and effects of instability in the world rice market. MSU International Development Paper No. 13. Michigan State University, East Lansing, Michigan
John A (2013) Price relations between export and domestic rice markets in Thailand. Food Policy 42:48–57
Nielsen CP (2003) Vietnam's rice policy: recent reforms and future opportunities. Asian Eco J 17:1–26
Perron P (1989) The great crash, the oil price shock and the unit root hypothesis. Econometrica 57:1361–1401
Petzel TE, Monke EA (1980) The integration of the international rice market. Food Res Inst Stud 17(3):307–326
Rapsomanikis G (2011) Price transmission and volatility spillovers in food markets. In: Prakash A (ed) Safeguarding food security in volatile global markets. Food and Agriculture Organization of the United Nations, Rome, pp 149–168
Rapsomanikis G, Hallam D, Conforti P (2003) Market integration and price transmission in selected food and cash crop markets of developing countries: review and applications. In: Commodity market review 2003-2004. Commodities Trade Division, Food and Agriculture Organization of the United Nations, Rome, pp 51–75
Rastegari-Henneberry S (1985) The world rice market: Giannini Foundation Information Series No. 85-2. University of California Davis, Davis, California
Seck PA, Tollens E, Wopereis MCS, Diagne A, Bamba I (2010) Rising trends and variability of rice prices: threats and opportunities for sub-Saharan Africa. Food Policy 35:403–411
Siamwalla A, Haykin S (1983) The world rice market: structure, conduct and performance. Research: Report 39: International Food Policy Research Institute, Washington, DC
Stage O, Rekve P (1998) Food security and food self-sufficiency: the economic strategies of peasants in Eastern Ethiopia. Euro J Dev Res 10(1):189–200
Timmer PC (2010) Management of rice reserve stocks in Asia: analytical issues and country experience. In: Commodity market review 2009-2010. Food and Agriculture Organization of the United Nations, Rome, pp 88–120
Tomek WG, Robinson KL (1990) Agricultural Product Prices, 3rd edition. Cornell University Press, Ithaca, NY
USDA (2013) GAIN report, Thailand: grain and feed update. GAIN report number: TH3111, Bangkok

Wailes EJ (2002) Trade liberalization in rice. In: Kennedy PL (ed) Agricultural trade policies in the new millennium. Haworth Press, New York

Wailes EJ (2005) Rice: global trade, protectionist policies, and the impact of trade liberalization. In: Aksoy MA, Beghin JC (ed) Global agricultural trade and developing countries. The World Bank, Washington, DC, pp 177–194

Xuifang D, Dwyer W (2008) Rethinking China's domestic agriculture support measures under WTO protocols. J Asia Pacific Eco 13:82–98

12

Innovation and valorization in supply chain network

Luigi Cembalo

Correspondence: cembalo@unina.it
Department of Agriculture, AgEcon and Policy Group, University of Naples Federico II, Via Università, 96 80055 Portici – Napoli, Italy

Abstract

Organizations, institutions and new governance mechanisms are taking place in the domain of innovation and valorization of the agrifood sector. Though not new in the literature, this topic is capturing the interest of scholars. It refers to the analysis of new organizations and governance frameworks in the agribusiness. Since this subject needs a deeper analysis, a platform of scientific-based discussion has been built, namely through the eleventh conference of the Wageningen International Conference on Chain and Network Management (WICaNeM) that was held in Capri (Italy) in June 2014 thanks to joining forces of three universities: Wageningen (NL), Bonn (DE) and Naples Federico II (ITA). WICaNeM is a well-known conference concerning management, economics, innovation, valorization and organization of chains and networks in a Life Sciences context. Thanks to Agricultural and Food Economics journal the 2014 conference benefits of a special issue where a selection of five papers were selected. In this short paper a brief presentation of those papers is provided.

Keywords: WICaNeM; Governance mechanism; Agribusiness

In the last two decades agrifood sector has been facing new challenges due to many changes occurred in social and political domains: food safety concerns, policy reforms, social and economic turbulences, often resulting in price volatility, just to cite some. As a reaction of these challenges players of the agrifood sector have been re-orienting their business throughout a re-thinking of their relationships within supply chains. Furthermore, rural communities have been re-organizing themselves in an attempt to create value developing a system of alternative food communities (Migliore et al. 2014). New governance structures, organizations and institutions are emerging aiming at innovation and valorization of the agrifood sector. This phenomenon is still under construction and it is capturing the interest of scholars. However, there still is a need for more in-depth investigations either from a theoretical point of view, case study based approaches, and innovative research methodologies.

These topics are not new in the literature. They refer to the analysis of new organizations and governance frameworks in the agribusiness. Williamson (2003, p. 23) and Masten (2000) pointed out that the agrifood sector traditionally provides a sound example of puzzling phenomena able to challenge mainstream theory. With the objective to go into such governance structures, a broader and deeper analysis is needed leading to new theoretical and methodological achievements as well as provide a rich and largely unexplored area for application and refinement of theories (Masten 2000).

Examples in this domain are the increasing number of consumer-producer networks: Solidarity Purchase Groups, Maintien d'une Agriculture Paysanne, Farmers Markets, and Community Supported Agriculture (Migliore et al. 2015; Cembalo et al. 2013). All these are strategies to re-organize supply chains, mainly at local level, rely on issues such as trust, fairness and social capital (Toler et al. 2009; Pascucci 2010). At a global level, on the other side, new forms of vertical coordination, partnerships and alliances are emerging among heterogeneous stakeholders. Such partnerships increasingly involve, at different levels of the supply chain, NGOs, universities, government agencies, international organizations, agrifood companies, and investors in the attempt to share information and specific assets and to realize common investments (Dentoni and Peterson 2011).

Over the last two decades, a platform of scientific-based discussion has been built. In June 2014 the eleventh conference of the Wageningen International Conference on Chain and Network Management (WICaNeM) was held in Capri (Italy) joining forces of three universities: Wageningen (NL), Bonn (DE) and Naples Federico II (ITA). WICaNeM has become a well-known conference regarding the management, economics, innovation, valorization and organization of chains and networks in a Life Sciences context. The conference also enhances discussions focused on building an agenda from an academic and research perspective, as well as from a company and business one. Put differently, WICaNeM provides the opportunity for attendees to meet large international companies and innovative SME providing vast opportunities to companies wanting to develop their business on an international platform, as well as researchers who want to prepare EU-programs.

Thanks to Agricultural and Food Economics journal the 2014 WICaNeM benefits of a special issue where a selection of value papers, among the best ones presented, were selected. After a peer review process, five papers were accepted for publication with a common research outline concerning innovation and valorization of the supply chain network.

Report

In "Case study analysis on supplier commitment to added value agri-food supply chains in New Zealand", authors Nic J Lees and Peter Nuthall focus on what attracts suppliers to be committed to long-term relationships in New Zealand agri-food supply chains where suppliers are required to consistently deliver to high product specifications. With a case study based approach authors look at factors determining enduring supply chain relationships. One of the most interesting result is that suppliers are attracted by the supply chain under study due to external economic conditions such as price uncertainty. Moreover, suppliers benefit of a premium price and relationship quality.

A second team of researchers explores a radical innovation in the food and feed industry. Their article "The perfect storm of business venturing? The case of entomology-based venture creation", authors Stefano Pascucci, Domenico Dentoni and Dimitrios Mitsopoulos, takes the reader to the issue of how to create a profit making entomology-based company. Authors first discuss a new venture creation in the agrifood sector. They then analyze the challenges of doing that in presence of a radical innovation. Implementing a venture creation game experiment authors highlight challenges and opportunities of such a business underlying the crucial role of trust and cooperation.

Moving geographically and in perspective, Marilia B Bossle, Marcia D de Barcellos and Luciana M Vieira analyze the production and consumption of eco-innovation food in Brazil. In "Eco-innovative Food in Brazil: Perceptions from producers and consumers", authors focus on both supply and demand sides in depth to see companies' motivation to adopt eco-innovation strategies, and consumers' perception. Results add empirical evidence on how values and general attitudes influence behavior towards eco-innovative food in the Brazilian food consumption context.

In "Who likes it sparkling? An empirical analysis of Prosecco consumers' profile", authors Laura Onofri, Vasco Boatto and Andrea Dal Bianco shift on the concept of certification of quality in a specific Region of Italy where Prosecco wine is produced. This wine can be certified as Controlled Denomination of Origin (CDO) and Controlled and Guaranteed Denomination of Origin (CGDO). The two certifications require different effort due to diverse quality standards, being CGDO stricter than CDO. However, most consumers seem not to make any difference between the two. For that reason authors made a study to elicit the preferences of consumers buying Prosecco in an attempt to understand how to add value to quality certification effort. A probit model was implemented on a large sample of homescan data.

A second team of Italian reasearchers explores the social embeddedness of a specific food community network. In "Alternative food networks as a way to embed mountain agriculture in the urban market: the case of Trentino", authors Emanuele Blasi, Clara Cicatiello, Barbara Pancino and Silvio Franco explore the mechanisms to deliver environmental and recreational value of peri-urban agriculture to the city. The role of traditional agriculture in peri-urban context risks to be replaced by non-agricultural and post-productive activities. So, farms re-organize themselves to move forward a multifunctional approach. One of the multifunctionality is the creation of alternative food network. The analysis is conducted in Trentino, Italy. The main result is the reinforcement of a link between the mountains and the city. Consumers' attitudes toward these practices are already very positive, thanks to their strong territorial identity. Results are in line with many other studies where it is shown the value creation of alternative food networks and their role in terms of social embeddedness.

Conclusions

Taken together, this collection of articles provides fresh insights into the multifaceted aspects of innovation and valorization of the supply chain network. The works presented here underline growing movements to relocate and revise the control and methods of food production as well as the relevance of sustainable method of production.

Competing interests
The authors declare that they have no competing interests

Acknowledgments
Author would like to thank AFE journal that made possible to disseminate some five papers presented at the WICaNeM conference 2014. A special thank goes to all participant that contributed to a successful conference.

References

Cembalo L, Migliore G, Schifani G (2013) Sustainability and new models of consumption: the solidarity purchasing groups in Sicily. J Agric Environ Ethics 26(1):281–303

Dentoni D, Peterson HC (2011) Multi-stakeholder sustainability alliances: a signalling theory approach. Int Food Agribusiness Manage Rev 14(5):83–108

Masten SE (2000) Transaction-Cost Economics and the Organization of Agricultural Transactions. In: Baye MR (ed) Advances in Applied Microeconomics - Industrial Organization, p 193

Migliore G, Schifani G, Dara Guccione G, Cembalo L (2014) Food community networks as leverage for social embeddedness. J Agric Environ Ethics 27(4):549–567

Migliore G, Schifani G, Cembalo L (2015) Opening the black box of food quality in the short supply chain: effects of conventions of quality on consumer choice. Food Qual Preferences 39(1):141–146

Pascucci S (2010) Governance structure, perception and innovation in credence food transactions: the role of food community networks. Int J Food Syst Dynamics 1(3):224–236

Toler S, Briggeman BC, Lusk JL, Adams DC (2009) Fairness, farmers markets, and local production. Am J Agric Econ 91 (5):1272–1278

Williamson OE (2003). Transaction cost economics and agriculture - An excursion. In Multifunctionality Agriculture: A New Paradigm for European Agriculture and Rural Development. Edited by Huylenbroeck GV, Durant G. Ashgate Pub Ltd. Aldershot, UK

13

Who likes it "sparkling"? An empirical analysis of Prosecco consumers' profile

Laura Onofri[1*], Vasco Boatto[2] and Andrea Dal Bianco[2]

* Correspondence: lonofri@unive.it
[1]Department of Economics,
University Cà Foscari of Venice, S.
Giobbe 873, 30121 Venice Italy
Full list of author information is
available at the end of the article

Abstract

The purpose of the study is to understand the profile (if any) of the typical Prosecco wine consumer, for both Controlled Denomination of Origin (CDO) and Controlled and Guaranteed Denomination of Origin (CGDO) types, with a twofold objective. First, the study aims at contributing to the economics literature dealing with opening the "black box of preferences" and understanding consumers' behavior. Second, more practically, the study aims to advise producers on the design of more targeted industrial strategies and policies. Using Homescan data collected from large-scale retail trade transactions in the period 2009-2011, we adopt a probit model and test a set of simple relationships between the probability that Prosecco (in both Geographical Indications) is purchased and selected consumers' socio-economic characteristics and product attributes. The results allow us to draft a profile of the typical consumer of Prosecco. The Prosecco CDO consumer lives in the North of Italy, is wealthy, relatively young, lives in a small (2-3 people) household and reacts to price changes. In addition to the latter feature, the typical Prosecco CGDO consumer has a preference for selected brands and extra-dry wine taste. Marginal effects are computed and predict that a 1% increase in Prosecco CDO price will decrease the probability that a consumer purchases the product by 0.36%. In addition, a 1% increase in Prosecco CGDO price will decrease the probability that a consumer purchases the product by 0.26%. The different sensitivity to price changes is corroborated by the fact that Prosecco CGDO consumers express a preference for the product characteristics (brand and taste) and might be more "loyal to the product" than Prosecco CDO purchasers. Further research will broaden the scale of analysis and adopt multinomial probit models in order to simultaneously assess the profile of different consumers for other types of sparkling wines, including Champagne and Franciacorta.

Background

Prosecco is an Italian sparkling or semi-sparkling[a] white wine, made mainly from "Glera" grapes, and is currently the wine with the fastest growing demand worldwide[b]. Although the name is derived from that of the Italian village of Prosecco near Trieste, where the grape may have originated, Prosecco is mostly produced in the Veneto region, mainly in Treviso province. Although Glera has been cultivated around the Conegliano and Valdobbiadene hills since the 18th century, Prosecco's success has only begun in the last decades, since its mass production as sparkling and semi-sparkling wine. There are now more than 25,000 ha of Glera vineyards, and more than 350 million bottles are produced annually[c]. Prosecco is now universally recognized as a high quality sparkling wine[d], and exported all over the world, especially to the USA,

Germany and the United Kingdom[e]. This sparkling white wine, rich in freshness, flavors, and with a low alcohol content, is strongly liked by consumers and sales are continuously increasing both in Italy and worldwide. In addition, the Prosecco production method is relatively inexpensive[f] if compared to those products that are "perceived" as substitutes by consumers of sparkling white wines, spanning from Franciacorta to Cava, to Trento, and even, for particular demand segments, *Champagne*.

Prosecco wine can be differentiated into Prosecco Controlled Denomination of Origin (CDO), and the Prosecco Controlled and Guaranteed Denomination of Origin (CGDO), depending on the geographical area where the grapes are cultivated. It is worth noting that both CDO and CGDO wines come under the European DOP classification (Protected Designation of Origin), so outside Italy are hypothetically of the same quality level. The reasons for the presence of two similar products on the market are to be found in the production regulation change that took place a few years ago. In 2009, the strong demand for Prosecco wine led to a need to increase supply, attained with a new regulation that allows the expansion of the Prosecco Area. In fact, the historical area of production, formerly Prosecco CDO area, gained the CGDO qualification, the more prestigious appellation among all Italian Geographical Indications (GI). The CGDO, in comparison with the CDO, has a stricter production protocol, and the quality of each batch is compulsorily checked by a tasting commission before being commercialized. At the same time, an extended Prosecco CDO Area that included two regions and seven provinces was created, leading to a fast expansion of Glera cultivated surface[g] (nowadays the CDO Prosecco is the biggest Italian wine GI[h]). Consequently, in the period 2010–2013 there was a 35% increase in Prosecco supply, with an equal increase in demand.

As Prosecco demand grows several questions arise, with both theoretical and practical implications. Who buys this wine? Is it possible to tackle the profile of a typical consumer or, in more technical terms, is it possible to elicit the preferences structure of those who purchase the wine? Is there any difference between Prosecco CDO and CGDO consumers? Gaining insights into this issue can contribute to the economic literature debate on preferences' assessment and consumer behavior. We know from the theory that the preference structure of consumers drives the choice, but we know little on the ways consumers form their preferences and orient consumption. In addition, the study aims to be more than just an intellectual exercise, since it can provide insights to Prosecco producers for industrial and pricing strategies. In this perspective, using Homescan data, collected from *Large-Scale Retail Trade* (LSRT) transactions in Italy in the period 2009–2011 we adopt a probit model and test a set of simple relationships between the probability that Prosecco (in both Geographical Indications) is purchased and selected consumers' socioeconomic characteristics and product attributes. Homescan data are collected and provided by A.C. Nielsen. Homescan data are very informative since contain information on both product characteristics and consumers' information[i].

This paper is organized as follows: Section 2 describes the research motivations and provides a survey of the economic literature on consumers' behavior and preferences' assessment in the wine sector; section 3 describes the Homescan data used in this research and provides selected descriptive statistics and background information. Section 4 explains the modeling strategy and comments on the estimation results. Section 5 gives the conclusions.

Rationale and literature survey

Neoclassical economics suggests that individuals choose according to self-interest and constraints. As (Andreoni and Miller 2008; page) 15 highlight "At its weakest, self-interest only means that choices conform to some underlying preference ordering that is complete, reflexive and transitive, and, hence, some utility function can be used to describe behavior". Individual preferences represent a dimension of choice and are formed and ordered according to criteria that, though not disputed in the way they are ordered and differ across individuals (Stigler and Becker 1977), are studied by economists who are committed to opening up the "black box of preferences" (Arrow, 1951). If it is true that "the individual may order all social states by whatever standards he deems relevant" (Arrow 1951, p. 17), it is very difficult to empirically assess those preferences, for whatever good. In this perspective, a suggestion for preferences' elicitation comes from (Andreoni and Miller 2008; page 15): "... the assumption of self- interest does not tell us what variables are in that utility function. What does? Our methodology is that people themselves, through their actions, will do so". The challenging task aimed at understanding and eliciting preferences is applied, in this study, to the consumption of the Prosecco wine. Following Andreoni and Miller (2008), we study the Prosecco consumers' behavior in order to assess and elicit their preference structure for the good, and more in particular, in order to profile the "typical" (if any) Prosecco consumer's socio-economic characteristics.

The economic literature has addressed the study of wine consumers' preferences and behavior by making use of elicitation methodologies based on both "stated" and "revealed" preference methods. In the research stream of "stated" preference studies, Gil and Sánchez (1997) used a conjoint designed experiment to examine and compare wine attribute preferences within and between two different Spanish regions. They evaluated the importance of three attributes: price, region of origin and grape vintage year, finding origin to be the most important attribute, but with relevant differences in consumer behavior between regions, in particular in terms of price sensitivity. Similar results were found by Mtimet and Albisu (2006), and Veale and Quester (2009). Mtimet and Albisu assessed Spanish Denominations of Origin (DO) wine consumer behavior through a choice experiment technique. They estimated willingness to pay based on four attributes: DO, price, wine aging and grape variety. Their results showed the DO and wine aging to be the most important in the consumer buying decision, although with some differences between frequent and occasional consumers. Veale and Quester (2009) found price and origin to be the most important attributes influencing consumer quality perceptions. Lockshin et al. (2006) used a discrete choice experiment to show how relative purchase rates change as brand, region, price and any award are changed. In addition, they found a price-quality effect, where demand increases as price increases, then drops after a certain point. Thiene et al. (2013) explored the effect of inclusion of answers to attitudinal questions in a latent class regression model of stated willingness to pay (WTP) for Prosecco. They found a reasonable pattern of differences in WTP for Prosecco according to DO and the emergence of important ancillary indicators of taste differences for specialty wines. Somogyi et al. (2011) assessed the underlying motivations of Chinese wine consumption through quantitative focus groups, with participants divided into groups based on age and gender. Their main findings were that Chinese wine consumers are influenced by face and status. In addition, the notion of wine consumption for health-related purposes was uncovered and a linkage found with traditional Chinese medicine. Among the other variables examined in the understanding of wine consumption habits and consumer preference we can highlight type of aging (Pérez-Magariño et al. 2011), alcohol strength (Saliba et al., 2013), color and style (Bruwer and Buller, 2012), reputation (Caracciolo et al., 2013), country of origin (Balestrini and Gamble, 2006; Bruwer and Buller, 2012; Di Vita et al., 2014), type of bottle closure (Marin et al., 2007). sustainability logos (Ginon et al., 2014) and gender (Bruwer et al., 2011).

In the literature on revealed preference methods, as applied to wine consumption issues, Ashenfelter (2008) used hedonic analysis for estimating consumers' implicit prices, e.g. valuation of the wine characteristics and quality attributes. The author gathered auction data on Bordeaux wines, and along with weather data, used this to predict the prices and quality of the wine. Nerlove (1995) used data from the Swedish state-importer of alcohol. He estimated hedonic price functions and the own-price demand elasticity for wine in Sweden, arguing that the state importation of wines resulted in completely elastic, parallel supply of wines. In particular, implicit prices for quality attributes are determined not from a regression of variety price on a vector of quality attributes, but rather from a regression of quantity sold (adjusted for weeks of availability) on price and quality attributes. Such a reduced form is justified by the assumption that prices and attribute contents can be taken as exogenous to the Swedish consumers, who are highly sensitive to price. Estimates of the implicit valuations of quality attributes are shown to differ greatly from those obtained from the more usual hedonic regression with price as the dependent variable. Combris et al. (1997) applied the hedonic price technique to Bordeaux wine. In the hedonic function, the authors included not only the objective characteristics appearing on the bottle label but also the sensory characteristics of the wine. Their data came from an experimental study in which juries evaluated and graded a sample of Bordeaux wines. The estimation of the hedonic price equation showed that the market price is essentially determined by the objective characteristics. The estimation of a jury grade equation showed that quality, unlike market price, is essentially determined by the sensory characteristics.

Considering the application of revealed preference methods for assessment of consumers' behavior and preferences specifically in the Italian wine sector, the following papers can be highlighted. Torrisi et al. (2006) used a linear almost ideal system to provide price and expenditure elasticities of Italian red table wine demand, finding a tendency to substitution across brands and a degree of competition among leading brands. Stasi et al. (2011) adopted quadratic almost ideal demand on a four equation system (QUAIDS) for estimating demand and elasticities (own-price and substitution) in order to test this hypothesis and verify the importance of DO in consumers' choice of wine. Estimates proved the existence of a differentiation effect of GIs (geographical indications) in terms of magnitude of elasticities and substitution effects. GIs corresponding to higher quality generate lower price sensitiveness and product substitution than wine without GIs. Controlled Origin Denomination (DOC) wine demand results are price sensitive and substitute for wines of different GIs. Controlled and Guaranteed Origin Denomination (DOCG) is the most profitable GIs. In fact, because of its inelastic demand, DOCG price could potentially be increased, to a certain extent, without any significant effect on volumes consumed. Cembalo et al. (2014) estimated a demand system (censored QUAIDS), using a statistically representative panel of 6,773 Italian households, to see to what extent, if any, substitution occurs in home consumption of basic wines, which is the main channel of distribution of inexpensive wines in Italy. The authors highlight the importance of packaging, such as a carton as an alternative to glass, in driving the preferences for cheap wines.

The present paper follows the literature on preference elicitation and understanding in wine markets. It is an attempt to understand consumers' behavior (and underlying preferences that drive choice) in the Italian Prosecco market. Differently from Thiene et al. (2013), the paper adopts a revealed preference method based on the empirical analysis of Homescan data[j] in order to understand what affects the choice to consume different types of Prosecco and what socio-economic and product characteristics determine the preferences for Prosecco. Specifically, we adopted a choice model derived in a random utility maximization model (RUM) framework, in which decision makers are assumed to be utility maximizers. The theoretical framework is based on Lancaster approach, asserting that a good per se does not give utility to the consumer. A good has a set of characteristics, and these characteristics may give rise to utility. In addition, Lancaster generalized that goods can posses multiple characteristics which can be shared by multiple goods separately (Lancaster, 1966). Following Lancaster, a consumer will choose the bundle of attributes of the goods that maximizes his/her utility to a budget constraint. Empirically, the relationship between products attributes/sociological variables and consumer preferences is formally investigated through a probit regression based on the RUM theory (Mc Fadden 2001). The empirical strategy differs from the above selected market valuation literature, because we do not model a hedonic price or a demand function, but attempt to elicit preferences by looking at a dichotomous behavior: the binary choice to purchase (or not) a certain type of Prosecco and the variables that affect the choice to purchase. The paper is a contribution to the attempt to understand consumers' behavior and the *"black box of preferences"* in the Prosecco market.

Methods

Data and empirical strategy

Data are provided by A.C. Nielsen. They are gathered from the wine purchase records, covering the period from January 2009 to December 2011, collected by A.C. Nielsen through scanner transactions in Italy, recorded by Homescan panelists at home[k]. The data cover 246,860 wine purchases, distributed over three years, made by 9,534 households and refer to 9,811 wine products sold in the LSRT. In particular, in this study we have downscaled the original large dataset by removing all transactions that did not refer to Prosecco wine. This means that we have downscaled the dataset to a total of 4,960 observations. The Prosecco dataset contains information about the selling price and purchased quantities, format and packaging[l], organoleptic characteristics[m] of the wines, geographical origin[n], brand[o], type of outlet[p] and location. The dataset also contains information about the panelists' socio-economic characteristics spanning from income, location, type of family and number of household members[q]. Table 1 describes the variables and summarizes the descriptive statistics of selected variables. The table contains three pieces of information: a) variables related to product characteristics; b) variables related to product marketing and c) variables related to consumers' socio-

Table 1 Descriptive statistics

Variable	Description	Mean (% frequency)	Std. Dev	Min	Max
Price	Price of Prosecco per liter (€)	7.02	2.36	0.2	25
Quantity	Liter of Prosecco purchased per person	1.02	1.33	0.4	22.5
Format	Prosecco j's bottle content in liters	0.75	0.07	0.2	1.5
Type	Brut	11%			
	Extra dry	37%			
	Dry	20%			
	Sweet	1%			
	Other	31%			
Denomination	CDO	65%			
	CGDO	23%			
	IGT	12%	*Produced before 2010*		
Seller type	Discount	9%			
	Hypermarket	45%			
	Supermarket	41%			
	Other	5%			
Household income	Low income	18%	*<535 € per capita per month*		
	Medium-low income	25%	*535 – 908 € per capita per month*		
	Medium-high income	32%	*908 – 1389 € per capita per month*		
	High income	25%	*>1389 € per capita per month*		
Household members	1 member	9.2%			
	2 members	30.1%			
	3 members	29.6%			
	4 members	24.4%			
	5+ members	6.6%			
Family organization	Pre families	2.8%			
	New families	10.8%			
	Established families	13.0%			
	Maturing families	11.9%			
	Post families	24.5%			
	Older couples	32.6%			
	Older singles	4.4%			
Consumers' age	<34 years	4.9%			
	35 – 44 years	24.2%			
	45 – 54 years	33.1%			
	55-54 years	21.9%			
	>65 years	15.9%			

Source: own elaboration from A.C. Nielsen Homescan data.

economic characteristics. In particular, selected descriptive statistics highlight that the wine is mainly purchased by people with a household income above average (32%); there is also relevant participation for households with an income both high (25%) and below average (25%), while purchases fall for those with low income (18%).

Table 2 Observations distribution over time

	Total	Prosecco CDO	Prosecco CGDO	CDO/CDOG ratio
2009	84,089	1,556	267	5.82
		(1.85%)	(0.32%)	
2010	82,935	1,640	270	6.07
		(1.98%)	(0.33%)	
2011	79,836	1,764	361	4.88
		(2.21%)	(0.45%)	

Source: own elaboration from A.C. Nielsen Homescan data.

In order to contextualize the study, we provide some information on the wine markets in Italy, looking at price and sales trends. Table 2 shows the relationship between the complete dataset transactions and those referred to Prosecco. It can be highlighted that whilst total wine purchases have decreased from 84,089 to 79,836, the purchase of Prosecco has risen from 1,556 to 1,764 for CDO type, and from 267 to 361 for CGDO, with an increase of the Prosecco share on total wine sales of 0.36% and 0.13%, respectively. In 2011 Prosecco represents 2.66% of the total wine sales made in the LSRT.

Table 3 reports a comparison among wine prices in the period 2009–2011. In 2011, the difference in unitary price between the two types of Prosecco was 1.07 €/L, while in 2009 and 2010 it was 2.22 and 2.96 €/L, respectively. Differently from other wines, Prosecco has experienced an inverse trend, showing an overall price increase of 21.4%. However, the price differential between Prosecco CDO and the average price of other sparkling wines, decreased from 4.24 €/L in 2009 to 1.01 €/L in 2011. The increase in the average price of Prosecco might depend on a different pricing strategy, aiming to equalize prices of Prosecco with the price of direct competitors like Asti and Franciacorta. In order to corroborate this interpretation, it is worth noting that the price of CGDO Prosecco, sold at a price similar to the Asti and Franciacorta prices, decreases by 7.2%. It is also interesting to note that in 2013 we can identify two different and well defined price clusters: one for sparkling wines, spanning from 6.06 to 7.24 €/L, and another for still and semi-sparkling wines, which spans from 2.26 to 2.91 €/L.

Table 3 price trends in LSRT (€/L)

	Average '09-'11	2009	2010	2011	Δ '09/'11
Prosecco	5.50	5.15	5.00	6.25	21.4%
CDO	5.16	4.72	4.60	6.06	28.4%
CGDO	7.38	7.68	7.42	7.13	−7.2%
Franciacorta	9.08	10.02	10.48	7.24	−27.7%
Asti	8.15	9.19	8.61	6.85	−25.5%
Lambrusco	3.57	4.05	3.84	2.91	−28.1%
Still red	3.94	4.50	4.36	2.95	−34.4%
Still white	3.38	3.68	3.58	2.93	−20.4%
Still rosè	3.11	3.57	3.43	2.26	−36.7%

Source: own elaboration from A.C. Nielsen Homescan data.

Table 4 Sales distribution of Prosecco and wine in the LSRT

	Prosecco	Wine
Discount	7.41%	11.03%
Hypermarket	47.57%	38.45%
Supermarket	39.71%	42.43%
LS	0.82%	2.23%
Free Service	0.67%	1.20%
Others	3.82%	4.66%

Source: own elaboration from A.C. Nielsen Homescan data.

Finally, Table 4 shows how sales are distributed, within the different LSRT channels, for Prosecco and other wine types.

Empirical strategy and estimation results

We wonder what spurs the consumer to choose a bottle of Prosecco CDO or Prosecco CGDO, and what types of product attributes and consumers' socio-economic characteristics affect that choice. We model the choice of purchasing Prosecco CDO (Prosecco CGDO) as a dichotomous choice. Each consumer is confronted with the (binary) choice to buy or not to buy the selected Prosecco type. The choice, is in turn, affected by a set of product characteristics and consumers' socio-economic characteristics. In

Table 5 Selected ML estimates. Probit Model

Explanatory Variables	(1) Consumption of Prosecco CDO	(2) Consumption of Prosecco CGDO
High Income	0.02*	-
Low Income	−0.19**	-
Pre Families	0.27	-
New Families	0.21*	0.31***
Maturing Families	−0.14	-
Post Families	0.36***	-
Older Couples	−0.05*	-
(Log)Price	−0.30***	−1.06***
(Log)Quantity	0.44***	0.91***
Age 35-44	0.25*	0.28*
Age 45-54	−0.29*	-
One household member	−0.45***	-
Two household members	0.20***	-
Three household members	−0.34*	-
Discount Market	−1.15***	-
Hypermarket	0.23***	0.24*
Carpené Malvolti	-	3.41***
La Gioiosa	-	1.59***
Extra-dry	-	0.34***
Constant	−0.38	−3.58*
R-squared	0.45	0.51

*** = 1% statistically significant; ** = 5% statistically significant; * = 10% statistically significant.

order to model such a dichotomous choice behavior, a linear regression model is generally inappropriate because this implies that the variance of the error term is not constant but dependent upon the explanatory variables and model parameters (see Veerbek 2000). To overcome the problems with a linear model, there is a class of binary choice models (or univariate dichotomous models), designed to model the choice between two discrete alternatives. A (general) relationship of this type can be modeled as follows:

$$P\{y_i = 1|x_i\} = G(x_i, \beta) \tag{1}$$

for some functions G(.). Equation (1) says that the probability of having $y_i = 1$ (the purchase of Prosecco CDO and CGDO) depends on the vector x_i, containing characteristics and variables that positively or negatively affect that probability[r]. The probit model described in Equation (1) is then estimated by maximum likelihood. We estimate, therefore, a set of simple relationships between the probability that Prosecco (CDO or CGDO) is purchased and some explanatory variables, including socioeconomic characteristics of the consumers (e.g. age, household type, income level) and product characteristics (e.g. price, brand, type etc.). Selected results are reported in Table 3, 4 and 5, column one for Prosecco CDO and column two for Prosecco CGDO.

The probability that consumers buy Prosecco CDO positively depends on several explanatory variables, for instance the fact that consumers belong to "post families"; are aged between 35–45, earn a high income and live in a small (two people) household. In addition, the probability that the consumers buy Prosecco CDO is positively affected by the purchase of the product at a hypermarket. On the contrary, the probability that the consumers buy Prosecco CDO, negatively depends on the fact that consumers belong to maturing families and/or older couples; are aged between 45–54; belong to a segment of low income earners and live in households composed of three members or more. If the purchase of the product is done at a discount market and the product price increases the probability of purchasing Prosecco CDO decreases.

For the Prosecco CGDO, the probability that consumers buy the product positively depends on several indicators, for instance, the fact that consumers belong to "new families" and are aged between 35–45 and that the purchase of the product is made at a hypermarket. For this type of Prosecco, brands matter and the probability that the product is extra-dry and branded Carpené Malvolti and La Gioiosa positively affects the purchase. The probability that the consumers buy Prosecco CGDO negatively depends on the increase of the product price.

If the description of the results is straightforward and follows the reading of the econometric estimates, their interpretation appears to be more challenging. Our results suggest that Prosecco is a product preferred by young, probably DINKs (double income, no kids) consumers, living in small households. The *Charmat* method confers a very light and fresh flavor to the wine, much appreciated by young people. The consumption characteristics of Prosecco CDO are very versatile, since the use of this wine spans from informal family and friends' gatherings to consumption at restaurants/pubs, to more formal occasions. What is preferred seems to be the possibility of an easy consumption: light wine, fresh flavor at low price. At the same time, the consumers react to price changes (price changes of a product that is not too costly), since an increase in the product price, as in the case of every normal good, negatively affects the probability of purchasing the product. There seems to be an apparent "paradox" in the interpretation of the empirical estimates, since young DINKs, earning high income, are reactive to price fluctuations and purchase the product at hypermarkets, probably when shopping for other commodities of day-to-day use. The apparent puzzle might be solved by considering a wider consumption bundle and the propensity to purchase a wider set of different products by these kinds of consumers. Therefore, when purchasing Prosecco, the high income young consumer has a strong revealed preference for the product, expressed in the purchase itself. At the same time, the consumer reacts to price changes. Therefore we can assess that he is maximizing his utility given a budget constraint which is not binding in strictly monetary terms but in terms of relative prices. These results are in line with the findings of Thiene et al. (2013), where Prosecco price changes induced almost 50% of a sample of consumers to reconsider their purchasing choice.

The characteristic of Prosecco as a wine "that aggregates socially" (since it is typically consumed at parties, bars and restaurants, for dinners and aperitifs) supports our empirical findings. In fact, the probability that the wine is purchased is negative in the case that the consumer belongs to maturing families and older couples. On the contrary, the probability is positive when consumers are young and living in small households. This is probably due to the fact that the socio-economic characteristics that negatively affect the purchase of Prosecco are related to traditional wine consumption paths. Older couples, indeed, more usually have a set of habits that include drinking still wine with their meals and conducting a quieter lifestyle. Younger consumers, on the contrary, are social creatures who enjoy the company of other people whilst drinking a fresh, light wine at a relatively low price. The socializing characteristics of Prosecco wine are appreciated beyond the original geographical area. In fact, northwestern Italy, traditionally characterized by a tasting preference towards strong ripened red wines, is nowadays the main Prosecco consumption pole. This is a clear sign of a change in consumers' taste.

The characteristics like brand and organoleptic attributes are appreciated only by Prosecco CGDO consumers. This is not surprising since the CGDO wineries are most prestigious, date back to the middle of the 20th century, and create a strong reputation associated to Prosecco wine year by year. In addition, if the CGDO wineries specialize in Prosecco production, the CDO ones produce a wide range of wines (still red, rosè and white, wine from dried grapes, etc.) and their name is generally less associated to Prosecco wine. The fact that the CGDO consumer pays more attention to organoleptic attributes could be due to his superior knowledge about Prosecco. It is likely that people willing to pay more and who choose historical Prosecco wineries have a better knowledge about Prosecco characteristics, and consequently prefer the organoleptic attributes reported on the label.

Finally, the interpretation of the results might suggest concrete indications for the design of the industrial policy of the Prosecco producer. It is important to highlight that the variables related to product differentiation (spanning from brand, to organoleptic, to bottle format) present estimated coefficients that are not statistically significant. Prosecco CDO is fresh, cheap and with a low alcohol content. These characteristics render the young DINKs the ideal consumers for this product. The main suggestion, therefore, is to use a simple pricing strategy. In fact we can corroborate this industrial policy suggestion by looking at the price trends of Prosecco with respect to other sparkling wines that could be considered substitutes for this product. As shown in section 3, the Prosecco price has increased over the studied period in order to align with the prices of the main substitutes. The competitive advantage of Prosecco relies in the fact that the *Charmat* method is relative cheaper with respect to the *Champenoise*, with which the other main Italian sparkling dry wines are produced. This represents an important insight for industrial policy since producers have to be very careful in their pricing strategy in order to avoid generating (marginal) losses of market shares. In particular, we can compute marginal effects and predict that a 1% increase in the Prosecco CDO price will decrease the probability that a consumer buys the product by 0.36%. In addition, a 1% increase in the Prosecco CGDO price will decrease the probability that a consumer buys the product by 0.26%. The different sensitivity to price changes is corroborated by the fact that Prosecco CGDO consumers express a preference for the product characteristics (brand and taste) and might be more loyal to the product than Prosecco CDO purchasers. The choice of the trade channel also impacts the pricing strategy. The LSRT can sell large stock of products at relatively low prices. This means that, especially for the smallest Prosecco wineries, it could be more convenient to look at other sales channels (e.g. *HoReCa*). This strategy seems to be already partially implemented by the firms, and is assessed in our data sample since only five brands (out of a total of 95) have sold almost 50% of the Prosecco[5].

Conclusions

Who likes it sparkling? A wealthy, relatively young Northern Italian, who lives in a small household. He/she buys Prosecco at a hypermarket and reacts to price changes at different rates: a 1% increase in the Prosecco CDO price will decrease the probability that he/she buys the product by 0.36%. In addition, a 1% increase in the Prosecco CGDO price will decrease the probability that he/she buys the product by 0.26%. He/she can choose among 95 brands, but the probability that he/she buys Prosecco CGDO increases if the brand is Carpené Malvolti or La Gioiosa. In addition he/she enjoys the extra-dry taste of the CGDO type. He/she is less interested in the brand nor in the taste when purchasing Prosecco CDO. On the contrary, those who do not like sparkling wine belong to maturing families and/or older couples, are aged between 45–54, belong to a segment of low income earners and live in a household composed of three members or more. The preference structure of Prosecco consumers is captured by the definition of the profile. A profiled consumer, characterized by selected socio-economic attributes prefers the product and selected characteristics of the product (selected brand, extra dry). The "black box of preferences" is (partially, within the limitation of the dataset), opened to show who has preference for what.

These are the bulk of the results derived from the present study that allowed us to highlight the profile of the Prosecco purchaser, therefore, to capture the latent preference structure for the product and the selected characteristics of the product. However, the results open a set of theoretical and empirical questions that remain unanswered for further research. First, if we were able to target who likes sparkling wine, we may be able to understand why they like it. *De gustubus disputandum non est* (Stiegler and

Becker, 1974), however, it is instructive and interesting to understand why a certain consumer profile is associated to the consumption pattern of a particular product. It is interesting to understand whether preferences are formed endogenously (depending on the personal, intimate nature of the individual/consumer) or if they are affected exogenously, with external (socio-economic and cultural) changes. In this perspective, for instance, consumers' preferences for wine in the last years have evolved towards lighter and fresher products, with lower alcohol content and easier to drink. An important, further research question aims at understanding whether this evidence is due to an exogenous effect on preferences' structures or is the result of an endogenous change in consumers' preferences. In order to address those points, further research will also aim at exploring different empirical strategies in order to simultaneously assess the profile of different consumers for different types of other wines, including champagnes and other sparkling wines.

Endnotes

[a]Still Prosecco has a very low incidence on total production.

[b]Unindustria convention, 29th October 2013.

[c]Data collected from "Rapporto di Distretto 2013", available at www.prosecco.it and from the "Bollettino del Consorzio di Tutela della Denominazione di Origine Controllata Prosecco", April 2014.

[d]Kinssies, Richard (July 10, 2002). "On Wine: Prosecco sparkle on their own terms". Seattle Post-Intelligencer. Retrieved 2008-12-29.

[e]CIRVE, 29th November 2013 – Presentation at Unindustria convention: USA, Germany and UK account together for more than 62% of total Prosecco imports.

[f]Sparkling and semi-sparkling Prosecco is produced with the *Charmat* method, an alternative method to the more "famous" *Champenoise* procedure for producing sparkling wines. The procedure presents economies of scale and allows the producers to market the product at low (average) prices.

[g]Source: Agenzia Veneta per i Pagamenti in Agricoltura (AVePA), 2014.

[h]The Prosecco Controlled Denomination of Origin (CDO), lies in an area that includes nine administrative provinces, and more than 600 municipalities, while the Prosecco Controlled and Guaranteed Denomination of Origin (CGDO) is produced in an area that includes 15 municipalities all in Treviso province.

[i]The Homescan panels are demographically representative of the household population and therefore the purchasing behavior of the panel can be grossed up to represent that of all households. Each household is equipped with a small handheld terminal through which details of all purchases are entered - product, quantity, price and outlet. This information, along with the date of purchase, is linked with demographic details of the household and the household purchasing history.

[j]Outside the wine sector, Homescan data have also been used to estimate brand level price elasticities and price response elasticities (Cotterill, 1994), change in household purchasing habits due to business cycle fluctuation (Cotti et al., 2014), the influence of selected demographic variables associated with purchase of organic milk (Alviola and Capps, 2010), the effect of taxes on sales (Harding et al., 2012) and the causes of price difference across households (Abe and Shiotani, 2014).

Table 6 Definition of socio-economic variables

Household affluence	Four groups of Households have been considered defined according to ranking of "revenue per consumption unit", proposed by OECD and calculated with the following:

$$Per\ capita\ income = \frac{Net\ Household\ Income}{1+0,7\ *\ (household\ size-number\ of\ children-1)+0,5\ *\ number\ of\ children}$$

Breakouts: (i) low affluence, 20%; (ii) below-average affluence 30%; (iii) above average affluence 30%; (iv) high affluence, 20%.

Types of families	**Pre Families:**

– i) Households with one member under 35 years old.

– ii) Household with 2 or more members with the housewife aged under 35 years, and with no children under 18 years of age.

New Families:

– Households with children under 6 years of age only.

Maturing Families:

– Households with children aged 0–17 years, and not all aged less than 6 years, or all aged above 10 years (i.e. not in categories 2 or 4).

Established Families:

– Households with children aged 11–17 only.

Post Families:

– i) Households with one member aged between 35 and 54 years.

– ii) Household with 2 or more members, with the housewife aged between 35 and 54 years, and with no children under 18 years of age.

Older Couples:

– Household with 2 or more members, with the housewife aged 55+, and with no children under 18 years of age.

Older Singles:

– Households with one member aged 55+ years.

[k]Selected consumers have an agreement with A.C. Nielsen and record their purchases at home, with a special device procured by Nielsen. In this way, Nielsen can form a consistent database where the products, the characteristics of the products and the socio-economic characteristics are constantly recorded. Nielsen sells the database to whoever is willing to buy it.

[l]Glass, bag in box, plastic etc.

[m]Color, sugar content, aging etc.

[n]CGDO, CDO, GTI, without any IG.

[o]Overall 2,914 brands were collected in the dataset.

[p]Supermarket, Hypermarket, Discounts, LS, other.

[q]See Table 6

[r]The function G(.) should take on values in the interval (0, 1) only. Attention can be restricted to the function G(xi, β) = F(x'i, β). As F(.) also has to be 0 and 1, F(.) can be chosen as some distributional function. A common choice is the normal standard distribution function, leading to the probit model (see Verbeek).

[s]The Prosecco market has several peculiar features. First of all, at wholesale level, production is characterized by very low concentration of supply, due to the presence of many small-medium wineries with a small market share[s]. At retail level, however, the marketing strategy allows a few large brands to be leaders in the LSRT segment[s]. Other competitors generally choose different retail trade scales, preferring to market their

products through the *HoReCa* channel, or through specialized and selected wine shops. This can be explained by the fact that LSRT channel requires high supply potential at a relatively low price, and big companies can thus benefit from economies of scale and higher stock capacity. Instead, medium and small companies mainly try to allocate their products in the channels that allow a better unitary profit.

Competing interests
The authors declare that they have no competing interests.

Authors' contributions
Authors are equally responsible of every paragraph of the paper. All authors read and approved the final manuscript.

Author details
[1]Department of Economics, University Cà Foscari of Venice, S. Giobbe 873, 30121 Venice Italy. [2]Department of Land, Environment, Agricolture and Forestry (TESAF), University of Padua, viale dell'Università, 16, 35020 Legnaro PD, Italy.

References
Abe N, Shiotani K (2014) Who faces higher prices? An empirical analysis based on Japanese Homescan data. Asian Econ Pol Rev 9(1):94–115
Andreoni J, Miller J (2008) Analyzing choice with revealedpreference: is altruism rational? In: Plott, C., Smith, V. (Eds.), Handbook of Experimental Economics Results. vol 1. Elsevier, North Holland, ISBN-13: 978-0444826428, ISBN-10: 0444826424
Alviola PA, Capps O (2010) Household demand analysis of organic and conventional fluid milk in the United States based on the 2004 Nielsen Homescan panel. Agribusiness 26(3):369–388
Arrow K (1951) Social choice and individual values. Wiley, New York
Ashenfelter O (2008) Predicting the quality and prices of Bordeaux wine. Econ J 118(529):174–184
Balestrini P, Gamble P (2006) Country-of-origin effects on Chinese wine consumers. Br Food J 108(5):396–412
Bruwer J, Buller C (2012) Country-of-origin (COO) brand preferences and associated knowledge levels of Japanese wine consumers. Journal of Product and Brand Management 21(5):307–316
Bruwer J, Saliba A, Miller B (2011) Consumer behaviour and sensory preference differences: implications for wine product marketing. J Consum Market 28(1):5–18
Caracciolo F, Cembalo L, Pomarici E (2013) The hedonic price for an Italian grape variety. Ital J Food Sci 25(3):289–294
Cembalo L, Caracciolo F, Pomarici E (2014) Drinking cheaply, the demand for basic wine in Italy. Aust J Agr Resour Econ 58(3):374–391
Combris P, Lecocq S, Visser M (1997) Estimation for a hedonic price equation for Bordeaux wine: Does quality matter? Econ J 107(441):390–402
Cotterill RW (1994) Scanner data: New opportunities for demand and competitive strategy analysis. Agricultural and Resource Economics Review 23(2):125–39.
Cotti CD, Dunn RA, Tefft N (2014) The Great Recession and Consumer Demand for Alcohol: Dynamic Panel-Data Analysis of U.S. Households. Available at SSRN. http://ssrn.com/abstract=2184415
Di Vita G, Chinnici G, D'Amico M (2014) Clustering attitudes and behaviours of Italian wine consumers. Quality - Access to Success 15(1):54–61
Gil JM, Sánchez M (1997) Consumer preferences for wine attributes: a conjoint approach. Br Food J 99(1):3–11
Ginon E, Ared G, Laboissière LHEDS, Brouard J, Issanchou S, Deliza R (2014) Logos indicating environmental sustainability in wine production: An exploratory study on how do Burgundy wine consumers perceive them. Food Res Int 62:837–845
Harding M, Leibtag E, Lovenheim MF (2012) The heterogeneous geographic and socioeconomic incidence of cigarette taxes: Evidence from Nielsen Homescan data. Am Econ J Econ Po 39:169–198
Lancaster KJ (1966) A new approach to consumer theory. J Polit Econ 74(2):132-157.
Lockshin L, Jarvis W, d'Hauteville F, Perrouty JP (2006) Using simulations from discrete choice experiments to measure consumer sensitivity to brand, region, price, and awards in wine choice. Food Quality and Preference 17(3):166–178
Mc Fadden D (2001) Economic choices. Am Econ Rev 91(3):351–378
Mtimet N, Albisu LM (2006) Spanish wine consumer behavior: a choice experiment approach. Agribusiness 22(3):343–362
Nerlove M (1995) Hedonic price functions and the measurement of preferences: the estimation for a hedonic price equation for Bordeaux wine: does quality matter? Eur Econ Rev 39:1697–1716
Pérez-Magariño S, Ortega-Heras M, González-Sanjosé ML (2011) Wine consumption habits and consumer preferences between wines aged in barrels or with chips. J Sci Food Agric 91(5):943–949
Saliba AJ, Ovington LA, Moran CC (2013) Consumer demand for low-alcohol wine in an Australian sample. International Journal of Wine Research 5:1–8
Somogyi S, Li E, Johnson T, Bruwer J, Bastian S (2011) The underlying motivations of Chinese wine consumer behaviour. Asia Pacific Journal of Marketing and Logistics 23(4):473–485
Stasi A, Nardone G, Viscecchia R, Seccia A (2011) Italian wine demand and differentiation effect of geographical indications. International Journal of Wine Business Research 23(1):49–61
Stigler GJ, Becker GS (1977) De Gustibus Non Est Disputandum. Am Econ Rev 67(2):76–90
Thiene M, Galletto L, Scarpa R, Boatto V (2013) Determinant of WTP for Prosecco wine: a latent class regression with attitudinal responses. Br Food J 115(2):279–299

Torrisi F, Stefani G, Seghieri C (2006) Use of scanner data to analyze the table wine demand in the Italian major retailing trade. Agribusiness 3:391–404

Veale R, Quester P (2009) Do consumer expectations match experience? Predicting the influence of price and country of origin on perceptions of product quality. Int Bus Rev 18(2):134–144

Veerbek M (2000) A Guide to Modern Econometrics. John Wiley & Sons, Baffins Lane, Chichester, West Sussex PO19 1UD, England. ISBN-10: 0470517697, ISBN-13: 978-0470517697

Permissions

The contributors of this book come from diverse backgrounds, making this book a truly international effort. This book will bring forth new frontiers with its revolutionizing research information and detailed analysis of the nascent developments around the world.

We would like to thank all the contributing authors for lending their expertise to make the book truly unique. They have played a crucial role in the development of this book. Without their invaluable contributions this book wouldn't have been possible. They have made vital efforts to compile up to date information on the varied aspects of this subject to make this book a valuable addition to the collection of many professionals and students.

This book was conceptualized with the vision of imparting up-to-date information and advanced data in this field. To ensure the same, a matchless editorial board was set up. Every individual on the board went through rigorous rounds of assessment to prove their worth. After which they invested a large part of their time researching and compiling the most relevant data for our readers.

The editorial board has been involved in producing this book since its inception. They have spent rigorous hours researching and exploring the diverse topics which have resulted in the successful publishing of this book. They have passed on their knowledge of decades through this book. To expedite this challenging task, the publisher supported the team at every step. A small team of assistant editors was also appointed to further simplify the editing procedure and attain best results for the readers.

Apart from the editorial board, the designing team has also invested a significant amount of their time in understanding the subject and creating the most relevant covers. They scrutinized every image to scout for the most suitable representation of the subject and create an appropriate cover for the book.

The publishing team has been an ardent support to the editorial, designing and production team. Their endless efforts to recruit the best for this project, has resulted in the accomplishment of this book. They are a veteran in the field of academics and their pool of knowledge is as vast as their experience in printing. Their expertise and guidance has proved useful at every step. Their uncompromising quality standards have made this book an exceptional effort. Their encouragement from time to time has been an inspiration for everyone.

The publisher and the editorial board hope that this book will prove to be a valuable piece of knowledge for researchers, students, practitioners and scholars across the globe.

List of Contributors

Pascal L Ghazalian
Department of Economics ,University of Lethbridge, 4401 University Drive, Lethbridge, Alberta T1K 3M4, Canada

Christian Ahlers
School of International Studies, Technische Universität Dresden, Dresden, Germany

Udo Broll
Department of Business and Economics, School of International Studies, Technische Universität Dresden, Helmholtzstr. 10, 01069 Dresden, Germany

Bernhard Eckwert
Department of Economics, University of Bielefeld, Bielefeld, Germany

Adele Coppola
Department of Agriculture, Agricultural Economics and Policy group, University of Naples Federico II, Via Università 100, 80055, Portici (Na), Italy

Fabio Verneau
Department of Agriculture, Agricultural Economics and Policy group, University of Naples Federico II, Via Università 100, 80055, Portici (Na), Italy

Narayan P Khanal
Graduate School for International Development and Cooperation(IDEC), Hiroshima University, 1-5-1Kagamiyama, Hiroshima 739-8529, Japan

Keshav L Maharjan
Graduate School for International Development and Cooperation(IDEC), Hiroshima University, 1-5-1Kagamiyama, Hiroshima 739-8529, Japan

Emanuele Blasi
Department of Economics and Management, Università degli Studi della Tuscia, via del Paradiso 47, Viterbo 01100, Italy

Clara Cicatiello
Department of Economics and Management, Università degli Studi della Tuscia, via del Paradiso 47, Viterbo 01100, Italy

Barbara Pancino
Department of Economics and Management, Università degli Studi della Tuscia, via del Paradiso 47, Viterbo 01100, Italy

Silvio Franco
Department of Economics and Management, Università degli Studi della Tuscia, via del Paradiso 47, Viterbo 01100, Italy

Mansor H Ibrahim
International Center for Education in Islamic Finance (INCEIF), Lorong Universiti A, 59100 Kuala Lumpur, Malaysia

Matteo Aepli
Agricultural Economics Group, Institute for Environmental Decisions, ETH Zurich, Sonneggstrasse 33, SOL E5 8092 Zurich, Switzerland

Pei Xu
California State University at Fresno, Department of Agricultural Business, 5245 N Backer Avenue, M/S PB101, Fresno, CA 93740-8001, USA

Zhigang Wang
School of Agricultural Economics and Rural Development, Renmin University of China, Haidian, Beijing 100091, People's Republic of China

Nic J Lees
Faculty of Agribusiness and Commerce, Lincoln University, Ellesmere Junction Road, Christchurch 7647, New Zealand

Peter Nuthall
Faculty of Agribusiness and Commerce, Lincoln University, Ellesmere Junction Road, Christchurch 7647, New Zealand

Glynn T Tonsor
Department of Agriculture Economics Kansas State University, 342 Waters Hall, Manhattan ITS 66506, USA

Ted C Schroeder
Department of Agriculture Economics Kansas State University, 342 Waters Hall, Manhattan ITS 66506, USA

Adam John
Institute of Agricultural and Food Policy Studies, Universiti Putra Malaysia, Putra Infoport, Jalan Kajang-Puchong, 43400 UPM, Serdang, Selangor, Malaysia

Luigi Cembalo
Department of Agriculture, AgEcon and Policy Group, University of Naples Federico II, Via Università, 96 80055 Portici – Napoli, Italy

Laura Onofri
Department of Economics, University Cà Foscari of
Venice, S. Giobbe 873, 30121 Venice Italy

Vasco Boatto
Department of Land, Environment, Agricolture
and Forestry (TESAF), University of Padua, viale
dell'Università, 16, 35020 Legnaro PD, Italy

Andrea Dal Bianco
Department of Land, Environment, Agricolture
and Forestry (TESAF), University of Padua, viale
dell'Università, 16, 35020 Legnaro PD, Italy

* 9 7 8 1 6 8 2 8 6 0 4 7 2 *